Southern Space Studies

The Southern Space Studies series presents analyses of space trends, market evolutions, policies, strategies and regulations, as well as the related social, economic and political challenges of space-related activities in the Global South, with a particular focus on developing countries in Africa and Latin America. Obtaining inside information from emerging space-faring countries in these regions is pivotal to establish and strengthen efficient and beneficial cooperation mechanisms in the space arena, and to gain a deeper understanding of their rapidly evolving space activities. To this end, the series provides transdisciplinary information for a fruitful development of space activities in relevant countries and cooperation with established space-faring nations. It is, therefore, a reference compilation for space activities in these areas.

The volumes of the series are peer-reviewed.

Samuel Oyewole

Utilitarianism in Outer Space

Space Policy, Socioeconomic Development and Security Strategies in Nigeria and South Africa

Springer

Samuel Oyewole
Department of Political Science
University of Pretoria
Pretoria, South Africa

ISSN 2523-3718 ISSN 2523-3726 (electronic)
Southern Space Studies
ISBN 978-3-031-49645-5 ISBN 978-3-031-49646-2 (eBook)
https://doi.org/10.1007/978-3-031-49646-2

This Springer imprint is published by the registered company Springer Nature Switzerland AG
The registered company address is: Gewerbestrasse 11, 6330 Cham, Switzerland

Paper in this product is recyclable.

Preface

Space policy, especially in developing countries such as Nigeria and South Africa, is motivated by promises that space and the attendant technologies have utilitarian values, that is, the potentials to advance the socioeconomic development and security aspirations of the affected nations. Despite decades of orbital journey, however, academic research, and public interest and understanding remain inadequate on the space politics, policies, strategies, institutions, programmes, capacities and capabilities, and the extent that targeted goals and expected benefits have been realised in these countries. Transcending the pure and applied sciences reductionism of space studies in Africa, this book offers social science perspectives on this subject. Accordingly, this book x-ray the intricate relationships of historical, geographical, social, demographic, economic, political, administrative, and strategic factors (at the national, regional, and global levels) that have shaped research and development trajectories of space science and technologies in Nigeria and South Africa.

Beyond the attempt to determine the development and security interests and perspectives of Nigeria and South Africa in space, this book provides an extensive survey of civil, commercial, and military space programmes of these countries, with special attention for various institutions (governmental, non-governmental, and private entities) that are involved, the resources that are available to them (funds, human, and materials: facilities and equipment), and their capacities to mobilise and covert these into capabilities (space assets and access). It is against this background that this book assesses the progress and challenges of Nigeria and South Africa in advancing the development and security aspirations of their peoples and governments with space capabilities, notably in support of national competitiveness, economic growth and transformation, indigenous technological capacity, defence, security, power projection, disaster management, human rights and the struggle for freedom, democracy, effective and good governance, food and water security, sustainable environment, quality health and education systems, and international cooperation.

This book provides Nigerian and South African experiences on space politics, policy, and strategy in relations to national and human dimensions of development and security of the aforementioned countries, as a contribution to broader African and the Global-South perspectives on the subject.

Pretoria, South Africa Samuel Oyewole

Contents

About the Author

Dr. Samuel Oyewole is a lecturer at the Department of Political Science, Federal University Oye-Ekiti, Nigeria, and a post-doctoral research fellow at the Department of Political Sciences and the African Centre for the Study of the United States, University of Pretoria, South Africa. He obtained B.Sc. Political Science and International Relations from the Osun State University and M.Sc. Political Science from the University of Ilorin, both in Nigeria, and Ph.D. International Relations from the North-West University, South Africa. His research interests cover aerospace, defence and security studies, African affairs, development studies, disaster management, and international relations. His articles have appeared in many journals, including *Astropolitics, Australian Journal of Maritime and Ocean Affairs, African Security, African Security Review, African Conflict and Peacebuilding Review, Defense & Security Analysis, Defence Studies, Democracy and Security, Geojournal, Journal of Asian and African Studies, New Zealand International Review, Politikon, Politeia, RUSI journal, South African Journal of International Affairs, Strategic Analysis,* and *Studies in Conflict & Terrorism.* His co-edited books include *Power Politics in Africa: Nigeria and South Africa in Comparative Perspective* (Cambridge Scholar Publishing, 2020); *Boko Haram's Campaign of Terror in Nigeria: Context, Dimensions and Emerging Trajectories* (Routledge, 2021); *Introduction to Political Science: A Reading Text on Politics and Governance* (FUOYE, 2023); *Armed Banditry in Nigeria: Evolution, Dynamics and Trajectories* (Palgrave Macmillan, 2024); *The Political Economy of Kidnapping and Insecurity in Nigeria: Beyond News and Rumours* (Springer, 2024). Email: samueloyewole47@yahoo.co.uk.

Part I

Introduction

Introduction: On Space Policy, Development and Security in Nigeria and South Africa

1

Abstract

This chapter introduces space policy, socioeconomic development and security in Nigeria and South Africa, and locates these countries in theoretical and empirical discussions in space politics and policy studies. In this case, utilitarian theory is adopted as a framework of analysing space policy, with the desire to promote socioeconomic development and security of the affected parties. Utilitarianism is relevant to post-Cold War efforts by space nations, especially the emerging ones, such as Nigeria and South Africa, to transcend the struggle for power, influence, status and prestige for tangible development and security utilities in space. In addition, this chapter examines space politics and policy in Nigeria and South Africa under three levels of analysis. At the global level, these countries are emerging space players; at the regional level, they are leading space players; and at the national level, they are made up of competing and complementary space and related institutions and interests. At all these levels, this chapter identifies the state and inadequacies of the extant literature on the subject. It is against this background that this book examines the space policy interests and commitments of Nigeria and South Africa, their determinants, trajectories, operationalisations, and contributions to socioeconomic development and security of the countries.

1.1 Background

Space policy broadly covers all human interests and associated activities in outer space and the celestial bodies. The interests can be political, military, and socioeconomic in nature, and their projectors can be state or non-state actors, such as corporate bodies, international organisations, and civil society groups. Space policy entails a series of programmes and activities, including rocketry and space

S. Oyewole, *Utilitarianism in Outer Space*, Southern Space Studies,
https://doi.org/10.1007/978-3-031-49646-2_1

flights (either manned or unmanned), artificial satellite systems (for research, communication, navigation, and remote sensing), astronomy and cosmology, and all related research and development (R&D). Human understanding and belief system on space has evolved from simplistic assumption that equate it to heaven, the home of the gods and related metaphysical constructs, to emerging realities that defined it as a utilitarian frontier to advance the development and security interests of mankind and states with required capacity. Just like airspace, land, sea, and cyberspace domains, outer space is one of the available utilitarian frontiers for humanity and nation-state to advance socioeconomic and security interests.

Space has attracted international political attention since the advent of rocket flight, and particularly the long-range ballistic missiles (Van Riper 2004). However, the success of the Union of Soviet Socialist Republic (USSR) in placing the first artificial satellite of the Earth in orbit on October 4, 1957, marked the advent of space age. Thereafter, space exploration between 1957 and 1980s was dominated by the Cold War rivalry between the USSR and the United States of America (USA) (Dolman 2002; Sadeh 2002). Other notable players in space from the Cold War era are Britain, China, France, Japan, Italy, India, Israel, and Brazil (Harding 2013; Sadeh 2013). In addition, there are emerging space players, such as South Africa, Pakistan, South Korea, North Korea, Egypt, Iran, Iraq, Argentina, Algeria, Indonesia, Taiwan, Turkey, and Nigeria among others (Harding 2013; Lal et al. 2015; Mistry and Gopalaswamy 2012; Oyewole 2017, 2020a; Wood and Weigel 2012). In 2017, Satellite Database of the Union of Concern Scientists (UCS 2020) documented no fewer than 56 countries that owned at least a satellite in orbit. Euroconsult (2017) equally recognised that 70 countries invested in space in 2016, of which more than 60 of them spent more than US$10 million and seven of them (the USA, China, Russia, Japan, France, Germany, and India) spent above US$1 billion. At the end of 2020, the United Nations Office for Outer Space Affairs (UNOOSA 2021) recognised that more 80 countries have launched a satellite into orbit.

Space activities involved limited number of actors during the Cold War, when space politics was dominated by competition for power and prestige, and space policy was militarised and directed toward state-centric security interests (Dolman 2002; Harding 2013; Sadeh 2013). Today, there is a growing understanding that the expanding number of countries that are involved in space is contributing to the democratisation of outer space and diversification of the motivating interests (Harding 2016:1). The interests in space have been renewed globally since the end of Cold War, and in developing regions particularly, on the basis that it holds significant prospects for security and development aspirations of the concerned peoples and governments (Argoun 2012; Harding 2013). In this case, satellite and its widely consumed applications are key to space popularisation. Satellites are employed for different purposes based on their designs. A communication satellite supports tele-communication, tele-education, tele-medicine, e-commerce and e-governance, among other applications. Earth observatory satellite supports remote sensing for weather forecast, precision agriculture, land and ocean survey, health survey, urban and regional planning, resource mapping, early warning and disaster management (Dick 2015; Maini and Agrawal 2014; Olla 2009; Pelton et al. 2017). Military powers also prioritised space based or supported command, control, communications, computers, intelligence, surveillance, and reconnaissance (C4ISR) in defence and security (HDoA 2005; Klein 2006; Preston et al. 2002). Beyond power and prestige, these civil and military utilities of space and attendant technologies are central to the aspirations and strategies of most space nations, especially the emerging ones, after the Cold War.

Despite the progress recorded by a growing number of developing countries in space, there is limited attention to their experiences and perspective in space policy studies. Amidst these, the experiences of space players in Africa are the least heard. A report by African Association of Remote Sensing of the Environment (AARSE) and European Association of Remote Sensing Companies (EARSC) (2016:28) revealed that there are more than twenty national remote sensing agencies and regional centres, and over 45 universities dedicated to building expertise in Earth Observatory (EO) technologies across Africa. The revenue of African space industry, which was made up of 283 companies across 31 countries, was USD7.37 billion in 2019 and projected to be USD 10.24 billion by 2024 (Space in Africa 2021a, 2021b). However, African space economy was already valued at USD 19.49 billion in 2021 and projected to be USD 22.64 billion by 2026 (Space in Africa 2022). As of early 2023, about 30 African countries have initiated space programmes, of which 15 of them owned at least a satellite in space. In total, 57 African satellites have been launched as of April 2023 (Oyewole 2023a). At the same time, no fewer than four African countries (South Africa, Egypt, Morocco, and Nigeria) have developed military space strategy and launch dual or military satellite (Oyewole 2023a, 2023b). These indicate the growing development of civil, commercial and military space strategies across Africa.

Although they remain generally limited, there are emerging interests in African space strategies in the framework of Southern Space discourse. This development has generated a series of continental surveys and brief multiple-countries-studies of African space strategies (Froehlich 2019; Froehlich and Siebrits 2019; Froehlich et al. 2020; Harding 2016; Oyewole 2017, 2020a). In addition to these, limited case studies and comparative analysis have the prospects of deepening the understanding and exposing the uniqueness of different space players in the region. Accordingly, this book explores space policy in the utilitarian framework of development and security of Nigeria and South Africa. Subsequent sections of this chapter provide the utilitarian framework of the analysis, brief background information on space policy of the selected countries, relevant levels of analysis, coupled with the objectives and organisation of this book.

1.2 Utilitarian Framework of Space Policy Analysis

Utilitarian theory is a veritable framework to assess the action, decision, or policy programme of any actor whatsoever, no matter their interest, in almost if not in all situations. This include the policy interests and activities or programmes of any state (indiscriminate of their level of development or power capabilities) and non-state actors (indiscriminate of their individuality or organisational nature as well as their locality or internationality) in space. This is unlike any other major theories of International Relations. For instance, realism only focus on state as the central player in space, reduces its interests to defence, security, and power projection, and tends to foreclose the prospects of international cooperation (Klein 2006; Moltz 2014; Mowthorpe 2004). Equally, liberalism focuses on the rights and freedom of all states and peoples to explore space peacefully in cooperation for the benefits of all mankind (Wolter 2006; Hudgins 2002; UN 2002); taking for granted the inequality in access to space, its technologies, and associated benefits. While social constructivism creditably offers multiple perspectives on actors and their interests in International Relations and by extention in space, it remains largely advocatory with little or no concise parameter for a study that seeks to assess socioeconomic and security benefits of space policy.

Utilitarianism is deeply rooted in the works of ancient moralists and philosophers like Mo Tzu (around 481 to 221 BC), Jesus Christ, Plato, Aristotle, Epicurus (341 to 271 BC), French Jean François, Marquis de Chastellux and Claude Helvétius, Scottish Francis Hutcheson (1694–1746), English David Hume (1711–76), Joseph Priestley (1733–1804), Adam Smith, William Paley (1743–1805), William Godwin (1756–1836), Jeremy Bentham (1748–1832) and John Stuart Mill (1806–73), Herbert Spencer and Henry Sidgwic (Scarre 1996; Mulgan 2007; Schultz 2017). Prominent in the new wave of utilitarian scholars are Green, Hobhouse, Ritchie and Hobson (Bayles 1968; Weinstein 2007). Amidst these, Jeremy Bentham and John Stuart Mill are leading proponents that articulate the theory. Although the theory and its principles remain a subject of debate in philosophy classes, they are widely applied, borrowed, and employed across all fields in social sciences and humanities.

Bentham (2000:14) argued that the principle of utility is a universal notion, which approves or disapproves every action whatsoever (by private individual, group or government), according to the tendency that it appears to have augment or diminish, promote or oppose the happiness of the party whose interest is in question. Utility is the property in any object, action or policy that tends to produce benefit, advantage, pleasure, good, or happiness, or prevent the happening of mischief, pain, evil, or unhappiness to the party whose interest is considered, individually or community generally (Bentham 2000:15–16). It is ethical, moral, just and rationally sound to direct the actions of an individual, a group or government to the production of the greatest possible quality and quantity of happiness of the one or those concerned and whose interests are in view (Bentham 2000; Mill 2001; Ng and Ho 2006; Sen and Williams 1992; Schultz 2017). Accordingly, it can be argued that utility is everything a state aspired to achieve in public policy, including in space.

There are multiple perspectives and variants of utilitarianism. Prominent among the variants are act-utilitarianism, rule-utilitarianism, and institutional utilitarianism. Act-utilitarianism is interested in the rightness or appropriateness of action or policy measure in accordance to how it maximises well-being (Mulgan 2007;

Scarre 1996). Nevertheless, there are those that believe that priority should be given to causal utility or the utility of motive (want, desire and intention) that give rise to an action in the final analysis of its utility (Adams 1990). Closely related to these is indirect utilitarianism, which also considered decision procedures (issues, such as the clarity or transparency, timeliness, friendliness, and resource management) as a priority in analysis (Mulgan 2007; Sen and Williams 1992). Rule-utilitarianism is premised on the notion that an action or policy can be said to conform to utilitarian principles when it follows moral, rational and widely accepted codes or rules that maximise well-being (Mulgan 2007; Lyons 1965; Bayles 1968). Institutional utilitarianism also argues that the best political, administrative, economic or social institutions, arrangements or regimes are those that maximise total welfare of those concerned. Beside these, there is consequentialism, which focus more on the outcome or effects of an action or policy, with priority for the one that best maximise the happiness or welfare of the party (Bentham 2000; Mulgan 2007; Sen and Williams 1992; Smart and Williams 1975). These will allow for critical analysis of space policy desires, intentions, motivations, institutions, actions, standards, regulations, processes, and consequences in selected cases, Nigeria and South Africa.

Utilitarianism assesses actions, policies and institutions based on their contribution to welfare or happiness of the affected parties. Economics often tend to operationalise happiness to mean standard of living, with priority for wealth production and distribution, of which per capital income is considered critical (Bruni and Porta 2007; Edward 1992). Generally, however, the term happiness means quality of life or wellbeing (Schultz 2017; Veenhoven 2004). This can be assessed using both qualitative and quantitative methods. From a political psychology perspective, it involves chances of good life or actual good life. Quality of life can mean liveability of an environment, life-ability of the person, utility of life, and satisfaction with life. Satisfaction can be of pleasure, part-satisfaction, top-experience and enduring or life-satisfaction (Bentham 2000; Veenhoven 2004; Ng and Ho 2006). Moreover, wellbeing can be subjective (mostly when its analysis is qualitative based) or objective (when is quantitative based) (Edward 1992; Mill 2001). These will aid critical assessment of space policy and its contribution to the advancement of the welfare, that is, socioeconomic development and security of the affected party—in this case, the peoples and governments of Nigeria and South Africa.

Despite the foregoing relevance of utilitarianism, there are many enduring contradictions and concerns against the theory. This theory recognised the right of every person to obtain benefits or happiness from the policy of their government as concerned parties/citizens. Yet, the same action that brought happiness or benefits to some parties can be disadvantageous to others, who may experience pain or pay the price for such action (Bentham 2000; Mill 2001). Many critics of utilitarian theory has described this as "taking from Paul to enrich Peter" and capitalised on this to question the ethical, rational, and just bases of its principles (Sen and Williams 1992; Smart and Williams 1975; Lyons 1965). The principle of utility, however, encouraged action with the tendency to promote the happiness of the community more than any it diminishes (Bentham 2000:15). Beyond rational appeal of different variants of utilitarianism that are well recognised, the logic of cost–benefit and causal-effect analyses tend to address this concern.

Other major concerns revolve around information, reductionism, and idealisation (Mulgan 2007; Sen and Williams 1992; Lyons 1965). Information is critical to utilitarian assessment, of which its inadequacy can undermine consequential analysis. The tendency to consider all situations, actors and interests as the same amount to reductionism. Idealisations involve preferences and choices, particularly on selection of what is allow to count and not (Sen and Williams 1992; Lyons 1965). However, most of these concerns are addressed in this study with the adoption of multiple sources of data, perspectives, and indicators, with mixed method in the analysis.

Against this background, utilitarian theory is a suitable framework to assess the contributions of space policy to development and security in Nigeria and South Africa. Harding (2013:5) is sensitive to the principle of utility when he argued that space policy involves conscious and specific use of space and its attendant technologies to promote the security and socioeconomic interests of a state. The relevance of this theory is further boosted by post-Cold War space politics, which prioritised realistic utilities (socioeconomic development and security) above idealistic ones (power, influence, status and prestige). For the growing number of developing countries in space, tangible utilities are even more important than scientific experiments and politics, especially to justify the cost. Moreover, utilitarianism is broad enough to accommodate different forms of utilities, which are associated with human and state-centric development and security interests. This framework is flexible to address as much as possible perspectives and indicators that are required for the assessment of space policy in emerging players such as Nigeria and South Africa and its contribution to development and security aspirations of their peoples and governments.

1.3 Nigeria and South Africa in Space

South Africa and Nigeria are leading space players in Sub-Saharan Africa, with records that are only matched by Algeria, Egypt and perhaps Morocco in North Africa. These are the first set of countries to develop space policy and establish agency for this purpose in Africa (Martinez 2016:38). Their space inventories also place them ahead of other countries in the continent. Between 1998 that Egypt launched the first African satellite and the end of 2016, when other countries in the region started to orbit satellite, the five countries had sponsored 25 satellites. The first satellites of these countries were placed in orbit between 1998 and 2003, making them the most experienced space players in the continent. As of April 2023, when 15 African countries own space asset, 67 per cent of all satellites (38 of 57) from the region were sponsored by Algeria (6), Egypt (13), Nigeria (7) and South Africa (12) (Oyewole 2023a). Amidst these countries, however, Nigeria has the most institutionalised space programmes, while South Africa has the most industrialised and liberalised programmes on the continent. Accordingly, this volume will explore the uniqueness of Nigeria and South Africa in space policy and politics as a contribution to African experiences and perspectives to the subject.

The South African National Space Agency (SANSA) was established in 2008 and national space policy was adopted in 2009. However, South African first satellite was launched into orbit in 1999. Unlike other space players in Africa, South Africa locally designed and produced most of its satellites (11 out of 12). Driving by military-oriented national security concerns, the apartheid regime in South Africa developed the most sophisticated space programme in Africa between 1960 and 1980s (Gottschalk 2010; van Wyk 2009). Its quest for nuclear weapons was also accompanied by policy commitments to develop missile and space launch capabilities and a satellite (Albright and Stricke, 2016; UNDDA 1991; Van Wyk 2014). While the satellite programme survived the country's political transition in 1990s, a considerable shift in policy focus from military-security to R&D objectives led to unilateral destruction of the nuclear weapons and suspension of the intermediate-range missile/orbital launch programme. With the launch of Condor E2 satellite in 2014, South Africa became the first country with a military satellite in Africa. Besides, South Africa can be described as the most advanced African nation in space science, given a host of space facilities, such as the Hartebeeshoek Radio Astronomy Observatory (HartRao), the South African Large Telescope (SALT), and the ongoing Square Kilometre Array (SKA) project. The country also possesses African most sophisticated industrial-base and private sector in space.

Nigeria declared interest in space during the Cold War, particularly in 1976, but committed little policy measures to this end until late 1990s (Abiodun 2017; James et al. 2014). In 1999, Nigeria established the National Space Research and Development Agency (NASRDA), the first of its kind in Africa. Again, in 2014, Nigeria became the first African country with institutionalised military space programme, as it established Defence Space Agency, which became known Defence Space Administration (DSA) in 2016. The country also established commercial space institutions, such as the Geo-Apps Plus Limited in 2003 and the Nigeria Communication Satellite (NIGCOMSAT) Limited in 2007. The Nigerian first satellite was launched in 2003 (Boroffice 2008). Ever since then, Nigeria has maintained an unbroken record of owning and operating active satellite in orbit, a record that is only matched in Africa by Algeria and Egypt. In 2007, Nigeria became the first country in the Sub-Saharan Africa to orbit a communication satellite. After South Africa, Nigeria became the second country in the continent to orbit a dedicated satellite for the military in 2023, considering that Egypt and Morocco only owned dual satellites. The Nigerian engineers have used British and Japanese facilities to design and develop two satellites. Nigeria is also developing its capacity in rocket flight and missile system (Fashade 2008; Oyewole 2020a), as well as astronomical facilities.

The space policy commitments of Nigeria and South Africa have reflected in their investments and spending on space programmes. The NASRDA took off with a budget of US$90 million in 1999 (Oyewole 2017). In 2013, Nigeria's space spending was US$70 million, which was more than US$68 million spent by the next top three in Sub-Saharan Africa (Angola, Congo, and South Africa) combined (Keith 2014). In 2014, space expenditure in Nigeria was US$66 million, and US$32 million in South Africa (Bochinger 2016). Amidst their various economic crises in 2016, Nigeria spent US$61 million and South Africa spent US$20 million on space programmes. Notably, Algeria, Morocco, and Angola outspent Nigeria in space, and in addition to the four countries, Egypt also outspent South Africa, in 2016 (Euroconsult 2017). Nigeria and South Africa led the continent in space expenditure in 2018, with US$61.7 million and US$60 million, respectively. After Algeria (US$100 million), South Africa (US$55.71 million) and Nigeria (US$50 million) were next African space spenders in 2019. These expenditures placed these African countries ahead of many space nations in Europe (such as, Australia, Belgium, Netherlands, Norway, Sweden, and Switzerland), Asia and Pacific (Australia, Israel, Iran, Pakistan, UAE, and Vietnam), and Latin America (Argentina and Brazil) (Space in Africa 2021a). During Covid-19 pandemic in 2020, South Africa spent US$168 million, one of its highest space spending, which accounted for 33.4 per cent of African total and placed it ahead of the next two leading African countries (Algeria and Nigeria) and the three leading Latin American and Carabian countries (Argentina, Bolivia, and Brazil) (Space in Africa 2021a).

Against this background, the policy priority, strategic drivers, and investment that space affairs are receiving in emerging space players like Nigeria and South Africa deserve better research attention. The pressing policy demand to advance the security and development aspirations of the peoples and governments of these countries here on Earth also necessitate this kind of study to interrogate their policy attention and investment in space. Beyond pure and applied science-reductionism of space research and development, therefore, the extent that space policy priority and investment of these developing countries have contributed to advance the aspirations of their peoples and governments deserve attention among social scientists. It is on these bases that this volume seeks to assess the space policy motivations, commitments and developments of Nigeria and South Africa, as well as the extent to which the identified policy objectives are operationalised and served by selected space programmes and investment.

1.4 Perspectives on Space Policy in Nigeria and South Africa

Space policy in Nigeria and South Africa can be considered from three major perspectives. These countries are emerging space players in the global context, leading African space players in regional context, and individually they are space nations with national context.

1.4.1 Global Level of Analysis: Nigeria and South Africa as Emerging Space Players

Nigeria and South Africa are emerging space players that often appear in most space policy studies with the goal of shifting academic attention on the subject matter from established space players to accommodate countries of the global-south (Argoun 2012; Harding 2013, 2016; Mistry and Gopalaswamy 2012; Wood and Weigel 2011, 2012). The end of Cold War and the growing accessibility and adoptability of small satellites in developing and emerging nations are changing the geopolitical dynamics of space in ways that cannot be ignored by space policy analysts. Notably, Harding (2013) provides an exceptional account of space policy in developing countries. The book described security and socio-economic development as universal interests of all space players, recognised the inadequate attention for the experiences of the developing countries, and then documented these for about 30 developing countries across the world. However, the study largely reduced socioeconomic development to economic growth, and security to narrow-minded military affairs. Some other studies have paid attention to the struggle for independent capacity in production and utilisation of satellites and in some cases launch vehicles as the core objectives of the developing nations in space (Argoun 2012; Mistry and Gopalaswamy 2012; Shabbir and Sarosh 2018; Wood and Weigel 2011, 2012).

Despite the growing number of policy studies on emerging space nations, Africa until recently, and Nigeria and South Africa particularly have received limited attention. Accordingly, the experiences and perspectives of space players in Africa are marginal in the field. For instance, a study on emerging space powers by Harvey et al. (2010) only focused on Asia, the Middle East, and South America. The limited and insubstantial reference to Africa and its space players in most of the yearbooks of space policy is another testimony of the position of the continent in knowledge production on the subject (Al-Ekabi and Ferretti 2018; Al-Ekabi et al. 2017; Burger and Bordacchini 2019). Equally, the celebrated developing countries-focused volume by Harding (2013) only recognised South Africa, Egypt, Nigeria, Algeria, and Tunisia among space players in Africa, and devoted 10 out of 236 pages to them. Some studies also focused on Algeria, Egypt, and Nigeria (Argoun 2012; Wood and Weigel 2012), while others included Kenya and South Africa to their limited list of emerging space nations from Africa (Wood and Weigel 2014).

Although many studies on emerging space nations have generated attention for Nigeria and South Africa, and Africa generally, in global context of space policy studies, they mostly lack depth. Most studies in this category often focus on very limited indicators of development and security, with little or no theoretical bases in policy studies, political economy, International Relations, and related disciplines in social sciences. Moreover, many of these studies lack harmony between conceptual and empirical details, which often fall short of comprehensive and systematic, and strategic or critical policy analysis. Some of these studies also recognised the connection between space applications and sustainable development, but failed to explore the progress of the selected countries in achieving the synergy in their space policies beyond paper works and political talks (Argoun 2012; Bhaskaranarayana et al. 2009). Hence, an in-depth study with limited cases of emerging African and other space players can contribute to scholarship in unravelling or deepening our knowledge of what Harding (2016) described as diverse interests that are associated with growing number of developing countries in space.

Against this background, studies on emerging space players have provided limited theoretical focus and empirical details of African experiences and perspective, which Nigeria and South Africa represent and have always championed. In as much as the Global-South or better still developing and emerging countries-oriented studies are essential for emancipation of Africa amidst Global-North or better still Western dominated knowledge production in the field of International Relations, the uniqueness in the continent's experiences and perspective cannot be totally ignored (Smith 2015; Ogunnubi and Oyewole 2020; Oyewole 2021). More important, very few Africans are involved in this category of research. Hence, the experiences of the continent and its leading space players such as Nigeria and South Africa are dominantly written by foreigners with their attendant value bias and interests (cf. de Klerk and Buchanan 2017). In some of these cases, therefore, the body of literature is largely a continuation of the dominant global narrative, the space popularisation projects of leading global players and the marginalisation of African experiences and perspectives.

1.4.2 Regional Level of Analysis: Nigeria and South Africa as African Leading Space Players

Nigeria and South Africa are regional powers in the continent (Amusan and Oyewole 2017; Oyewole 2020b). Hence, they have featured prominently in space studies in Africa as leading players in the continent. Although it remains considerably limited, there is a growing number of literatures on space and associated policy in Africa. In comparison with studies on global emerging space players, significant space studies on Africa are authored by Africans or those that are affiliated to African institutions (Abiodun 2008; Froehlich 2019; Froehlich and Siebrits 2019; Luncedo and Gottschalk 2013; Martinez 2012a, 2021b; Munsami and Offiong 2017; Oyewole 2017, 2020a). Based on a simple survey of major journals on space policy such as *Astropolitics*, *Space Policy*, and *Acta Astronomica*, it is evident that available studies on space policy and politics in Africa are dominantly carried out by Africans or those that are affiliated to African institutions. However, this cannot be taken too far, given the insignificant proportion of African studies in the general framework of space policy studies. Even in pure and applied sciences that tend to fair better, the proportion of space studies on Africa by Africans remains troubling. For instance, de Klerk and Buchanan (2017) surveyed the use of remote sensing for conservation and ecological studies in Africa and found 580 studies with authorship that spread across 98 countries. Amidst these, the USA came first on the list, as it accounted for 34% of the total publications, while South Africa came second with 22%. Besides, only seven African countries are listed in the top 20 countries on African-related space studies.

There is an emerging body of literature on space policy in Africa, covering a range of issues from regional to national space actors and associated dynamics. At the regional level, there are studies that situated Africa in the global arena or trend of space R&D, and largely provided a worrisome conclusion that the continent and its leading players lag behind their peers in policy priority and investment as well as advancement in space (Abiodun 2008). In contrast, some studies are considerably impressed and fascinated with the growing profile, visibility, and importance of the continent in space R&D as well as politics and policy (Gottschalk et al. 2018; Froehlich and Siebrits 2019; Oyewole 2017; 2023a; Space in Africa 2019). Other studies explored the emerging trend and prospects of regional space collaboration and institutions, such as the African Resource Management (ARM) constellation, the African Space Agency (ASA) and the African Leadership Conference on Space Science and Technology (Gottschalk 2008; Martinez 2012a; Mostert and Jacobs 2008; Munsami and Offiong 2017; Offiong and Munsami 2018). Notably, a study by Froehlich and Siebrits (2019) provides an extensive survey of international governmental organisations (IGOs) and the least explored international non-governmental organisations (INGOs) that are critical to regionalisation of space activities in Africa. This set of studies looks beyond state-centric reductionism of policy priority and investment as well as political commitments for sustainable space R&D for the people and government in Africa.

Regardless of the growing momentum for cooperation and regionalisation of space policy and programmes in Africa, the strength of nationalism that have underscored space programmes of key players across the continent have generated some concerns against regional integration and particularly the establishment of ASA (modelled after the European Space Agency (ESA)) (Martinez 2012b; Munsami and Offiong 2017). These, among others concerns, indicate the enduring relevance of regional powers or the key space players in emerging space policy analysis of the continent (Froehlich and Siebrits 2019; Oyewole 2017). In view of these, there are emerging body of literature that are counting and keeping scores or taking stocks of space assets of African countries in attempts to assess their capabilities as space players (Froehlich and Siebrits 2019; Luncedo and Gottschalk 2013; Oyewole 2017, 2020a, 2023b; Space in Africa 2019). However, some of these studies are more than mere stock taking.

Some studies have examined the level of international collaboration behind the capabilities of space players in Africa (Froehlich and Siebrits 2019; Oyewole 2017; Space in Africa 2019). Moreover, Oyewole (2017) assessed space policy priority and investment of African countries as well as associated cost and benefits. Cost was assessed in terms of funding by public and private actors locally, foreign direct investment (FDI) and priority amidst other pressing policy concerns, while benefits were considered in form of development in economic, education, governance, and security. However, the scope of the study is generally too broad and less in-depth. Besides, the study does not assess the achievement in space vis-à-vis designed strategic objectives of the affected countries. Froehlich and Siebrits's (2019) study filled some of these gaps by identifying socioeconomic objectives that are strategic to most African countries and available space related means and capabilities to attain them. Nevertheless, the study shows very little empirical evidence of how space capabilities have advance some of the identified socioeconomic objectives of the region. This concern was subsequently addressed in the follow up volumes among other studies, where the role of space in the struggle to achieve sustainable human development, peace, security, and governance in the continent and beyond were examined (Froehlich 2019; Froehlich et al. 2020; Amusan and Oyewole 2023; Oyewole 2022a, 2022b).

Beyond civil space strategies, African commercial and military space strategies have also received limited research attention. Space in Africa (2019) offers relevant insight into the development of African commercial space strategies, as it provides a comprehensive survey of space industry and economy in the region. The report is relevant to understand the liberal conception of development as deregulation, privatisation, and economic growth. However, it offers limited insight into how space means have advanced desired policy ends of the region and affected countries. Oyewole's (2020a, 2022c, 2023b) studies also provide insights into military-security dimension of space strategies in Africa, drawing from the experiences leading regional players such as Algeria, Egypt, Morocco, Nigeria and South Africa. However, in-depth study is required to further expand the coast of knowledge on these and other subjects of interest in African space politics and policy.

In most cases, regional studies on space policy, its contribution to development and security in Africa are able to focus on multiple countries with limited thematic scope. Accordingly, case study analysis with limited geographical scope is required to provide in-depth and multidimensional assessment of the contribution of space policy to development and security in the continent.

1.4.3 National Level of Analysis: Nigeria and South Africa as Space Nations

There are few but increasingly growing body of literature that have treated Nigeria and South Africa as space nations in their own right. Studies of this nature are mostly case study analysis, and their authors are dominantly citizens at home or in diaspora, and most importantly affiliated to national institutions. This category of literature often focuses on historical and current dynamics of space policy institutions, operations, and strategies in Nigeria and South Africa. Notably, the focus of space studies in Nigeria and South Africa are largely dominated by operational dynamics, which are dominated by pure and applied scientists and engineers. In the case of Nigeria, there is a growing body of literature on the prospects, progress, operations, and challenges of space technology adoption, with significant attention on satellite for remote sensing (Boroffice and Agbaje 2006; Ojo 2017). Beside the basic space technology adoption, space studies in this category in South Africa also cover the design, R&D, and operationalisation of satellites produced in the country (Koekemoer et al. 1999; Mostert et al. 2008; Mostert and Koekemoer 1997; Schoonwinkel et al. 2003). However, most of these studies often pay little or no attention to social dynamics, and particularly institutional and strategic dimensions of the subject matter.

Most studies on space policy in Nigeria focused on space laws and institutions, their structures, history, and in some cases, policy ends they are designed to achieve, and mechanisms, projects and programmes initiated to attain them (Abiodun 2017; Akinyede and Boroffice 2013; Fashade 2008; Isoun and Isoun 2013; James et al. 2014; Mohammed 2011; Momoh 2008; Tella 2018). Amidst these, NASRDA and its affiliated centres as well as their activities have dominated academic attention. Thus, literature is largely government or state-centric, with little recognition of non-state actors (especially private firms and civil society groups), which are in cooperation, competition, and perhaps conflict with public institutions in space policy formulation and implementation, and technological adoption and marketing. This is partly due to the predominance of state institutions in space initiatives, activities, sector, and policy processes in Nigeria.

Space policy studies in South Africa have largely focused on space history, institutions, laws, policy strategies, capacity building as well as programmes and projects (Gottschalk 2010; Martinez 2009, 2016; Munsami 2014; SANSA 2014; Sekhula 2013; The Tauri Group 2011; van Wyk 2009). Amidst these, the role of non-state actors, especially private individuals, firms, and civil society groups, in South Africa's space policy, programmes/activities, initiatives, sector and related history are more developed and recognised than in the case of Nigeria, even though they remain underrepresented in extant literature (Gottschalk 2010; The Tauri Group 2011).

Despite the growing research attention on the two countries, there is shortage of literature on military and security-dimension of their space programmes. There is also inadequate attention on the extent that space policy and strategy of the countries have been achieved over the years. Besides, space studies have attracted limited research attention from social scientists, especially economists and political scientists, and relevant specialists in International Relations, policy and strategic studies, and public administration in the two countries, like elsewhere in the region.

Against this background, it can be observed that space policy studies are underdeveloped in Nigeria and South Africa, as pure and applied scientists and engineers remain the principal policy analysts, while social scientists have taken the back sits. It is not surprising, therefore, that the social dynamics, including political, economic, policy, strategic, administrative, and international determinants and implications of space activities in the counties under consideration have not received sufficient academic attention. Accordingly, most of the achievements, prospects, and challenges of space programmes in these countries have not been properly assessed in relation with the development and security aspirations of their peoples and governments. Amidst these, the experiences of Nigeria and South Africa, like other African countries, have been largely buried under the space popularisation projects of the advanced space players. Hence, space policy studies in this consideration are dominated by what ought to be instead of what is, that is, what these countries ought to benefit from space policy priority and investment based on the experiences of the advanced space players and not what they have actually benefited in their own right.

1.5 The Puzzle and Objectives of the Book

The South African National Space Policy (2008) identified that "the pursuit of space activities in South Africa is informed by the overarching principle that these contribute to the country's economic growth and social development." Equally, the National Space Policy and Programme (2001) identified that Nigeria need to possess the "technical know-how to participate actively and independently in space activities and as a service provider of these products and services that have helped to serve the social and economic needs of the country." These indicate that these countries have notable years of experience in applications of space technologies with record of benefits that deserve research attention. However, most of the existing studies on this subject largely focus on historical context, institutional development, and the prospects and (possible) challenges of space policy and programmes in Nigeria and South Africa (Boroffice 2008; Gottschalk 2010; James et al. 2014; Martinez 2016; van Wyk 2009). As such, the convergence and divergence of what Nigeria and South Africa intended to achieve through space policy and what they have actually achieved over the years have received inadequate academic attention of policy analysts and social scientists. This concern cannot be ignored with the growing space spending and investment in these countries, considering the possibility of rechannelling the policy attention and associated funding to other means of addressing pressing national development and security challenges.

Developing countries like Nigeria and South Africa embraced space policy based on development and security promises that it has the potentials to inspire or contribute to economic growth, poverty alleviation, job creation, quality education, sustainable infrastructure, food security, water management, environmental sustainability, quality of health system, law enforcement, management of political and criminal violence, military modernisation, and credible disaster management capability among others (Argoun 2012; Harding 2016; Olla 2009). However, the experiences and perspectives of most developing countries remain largely unaccounted for in extant literature. Hence, the rationales for space policy priority and spending in these and many other emerging players are entangled in the space popularisation projects of the advanced players, whose core aim is to increase support base for their programmes locally, and power, prestige, and commerce internationally (Detsis and Detsis 2013; Oikonomou 2017). Accordingly, an enquiry into the space policies of Nigeria and South Africa is an important step to assess the myths and realities of space popularisation, especially on the benefits in space spending in the emerging players. These are primary subjects of concern that this study seeks to investigate.

This study seeks to assess the space policy programmes of Nigeria and South Africa as well as their contributions to socio-economic development and security. This book seeks to achieve the following objectives: to determine the policy objectives of Nigeria and South Africa in space; to examine the development of space policy in Nigeria and South Africa; to assess the benefits of space policy for the enduring developmental aspirations of the peoples and governments of Nigeria and South Africa; to examine the security utilities of space policy in Nigeria and South Africa.

1.6 Outline of the Book

This book is divided into four sections of nine chapters. The first section, which is made of two chapters, provides empirical, theoretical and conceptual introductions to this volume. The first chapter provides introduction to space policy, socioeconomic development and security in Nigeria and South Africa from empirical and theoretical perspectives. The chapter locates the two countries in theoretical and empirical discussions in space politics and policy studies. For this purpose, utilitarianism is adopted and projected as a suitable theoretical framework of analysing space policy, considering its desire to promote development and security of affected party. In addition, this chapter review the existing body of knowledge on space policy in Nigeria and South Africa. This is used to explore three levels of analysis (global, regional and national levels), and the inadequacies of the extant literatures on this subject. This chapter, thereafter, present the research puzzle and objectives that underscore this volume on space politics and policy studies.

The chapter two of this book focuses on the conceptual framework of analysis. It offers debate and clarity on the use of key concepts, such as space, space policy, development and security. On space, the chapter provides insights on the debates between pure and applied scientists on one hand and humanities and social scientists on the other hand, the resultant emergence of geometric and social spaces, and perspectives on absolute, relative and relational spaces. This is important to reiterate the multidisciplinary perspectives that are required in this study. Although space (outer space) is generally geometric and specifically astronomical, it is interlocked with social spaces in the policy context. On this basis, this chapter align with utilitarian conception of space policy as the use of space and attendant technologies to promote the socioeconomic development and security of the affected party. Thereafter, the concepts of development and security were interrogated and then summarised as the aspirations of the affected party (people and government).

The second section of this book, which is made up of three chapters, examines space policy, socio-economic development and security in Nigeria. The first chapter of the section, which double as the third section of the book, focuses on space politics, policy and strategy in Nigeria. This chapter examines Nigeria's space policy and strategy, and their social, economic, political, historical and international contexts and determinants. It provides background information on Nigeria's geography, history, state, societies, economy, politics and security that are essential to understand its policy, strategy, capacities, benefits and challenges in space. The chapter also examines the political, historical and socio-economic trajectories of Nigeria's space policy and strategy from cultural astronomy to spectator at the dawn of space age and then active player in space research and development. Equally, the chapter reviews extant laws, policies and strategies of Nigeria to determine national interests and perspectives on space, science and technologies, socio-economic development, and security, as well as their interconnectivity.

The chapter four of this volume examines the development of Nigeria's space capacities and capabilities and their connectivity to stated policy and strategic interests. Nigeria established African first dedicated agencies for civil and military space programmes, and fully state-owned corporations for commercialisation of space, as well as numerous centres of excellence for space research and development. In addition, several private companies and non-governmental organisations (NGOs) with exclusive or inclusive interests in space have emerged in Nigeria. Accordingly, this chapter provides an overview of the formation, locations, functions, financial strength, coupled with quality and quantity of human resources, facilities, and assets of various governmental, non-governmental and private institutions that are involved in civil, commercial and military space programmes, as well as space advocacies and popularisation in the country.

The fifth chapter of this book examines how space policy has promoted national and human dimensions of development and security in Nigeria. It provides insights into the convergence and divergence of space capacities and capabilities (such as institutional, financial, human and material resources, and assets), and the realisation of set policy objectives. It explores records and gaps in space support for national and human dimensions of development and security, as it relates to national competitiveness, economic growth and transformation, local technological capacity, defence, security, power projection, disaster management, human rights and struggle for freedom, democracy, effective and good governance, food and water security, sustainable environment, quality of health and education systems, and international cooperation.

Between chapters six and eight, the third section of this volume examines space policy, socio-economic development and security in South Africa. In this case, the South Africa's space policy and strategy, and the attendant capacities and capabilities have evolved from centuries of the nation's scientific, technological, political and socioeconomic developments, and its defence, security and international engagements. Accordingly, the sixth chapter examines the national and international politics of South Africa's space policy and strategy with the attendant capacities and capabilities, considering the sociocultural, economic, historical and military contexts, determinants and interests. It reviews the trajectories of South Africa's space policy and strategy from the indigenous cultural astronomy and the settlers' modernisation of space R&D in precolonial and colonial eras to amateur rocketry, militarised apartheid nuclear, missile and space programmes, and post-apartheid new dawn. Against this background, this chapter assesses extant laws, policies and strategies of South Africa to determine its national interests and perspectives on space and related technologies, socioeconomic development, and security.

The seventh chapter surveys various institutions (governmental, non-governmental and private entities) that are involved in South Africa's space policy processes, the resources that are available to them (funds, human and materials: facilities and equipment), and their capacities to mobilise and covert these into capabilities (space assets and access). Beyond different state-owned entities for civil, commercial and military space programmes and NGOs, South Africa has the most deregulated African space industry with flourishing private sector. Building on its space heritage, the government has supported academic innovations in educational institutions, which has generated spin-off key companies for the realisation of the objectives of the nation in space. Hence, the knowledge of the South African space capacities and capabilities is crucial to the convergence and divergence of the national policy interests in space and their realisation.

Chapter eight is interested in the extent that South Africa has employed its space and related capacities and capabilities to advance the development and security interests of the government and people of the country. It examines the convergence and divergence of South Africa's policy interests in space and their realisation. Hence, this chapter x-ray the progress and challenges of South Africa in space support for national and human dimensions of development and security, including issues of national competitiveness, economic growth and transformation, indigenous technological capacity, defence, security, power projection, disaster management, human rights and the struggle for freedom, democracy, effective governance, food and water security, sustainable environment, quality health and education systems, and international cooperation.

The fourth section of the volume provides the final conclusion, which is made up the ninth chapter. Hence, chapter nine provides a concluding reflection on space policy, development and security in Nigeria and South Africa. It summarises major themes that were considered across chapters on the subject of discussion, from where cross-examination, comparative analysis and conclusion on the experiences and perspectives of Nigeria and South Africa were drawn. On these bases, it is observed that Nigeria's space policy revolves around government and academic institutions with limited private sector, while South Africa's model is driven by academic and private sector with government coordination. The chapter equally explores the state of knowledge on space policy and politics in these countries, as parts of the emerging African and Global-South studies, as well as the unique contribution of this volume to literature, and some relevant areas for future research.

References

AARSEs and EARSC. 2016. Copernicus User Uptake: Inventory of Private Sector Companies in Earth Observation and Geospatial Fields in Africa. Draft Final Report for the European Commission, prepared by African Association of Remote Sensing of the Environment (AARSE) and European Association of Remote Sensing Companies (EARSC), with support from SpaceTec Partners.

Abiodun, A.A. 2008. Future trends in the global arena: Challenges and opportunities for Africa. *African Skies* 12: 12–22.

Abiodun, A.A. 2017. *Nigeria's Space Journey: Understanding its past, reshaping its future*. Abuja: African Space Foundation.

Akinyede, J.O., and R.A. Boroffice. 2013. *Nigeria's Quest in Space*. Ile-Ife: Obafemi Awolowo University Press.

Albright, D., and A. Stricker. 2016. *Revisiting South Africa's Nuclear Weapons Program: Its History, Dismantlement, and Lessons for Today*. Washington, DC: Institute for Science and International Security (ISIS) Press.

Al-Ekabi, C., and S. Ferretti, eds. 2018. *Yearbook on Space Policy 2016: Space for Sustainable Development*. Geneva: Springer International Publishing AG.

Al-Ekabi, C., B. Baranes, P. Hulsroj, and A. Lahcen, eds. 2017. *Yearbook on Space Policy 2015: Access to Space and the Evolution of Space Activities*. Geneva: Springer.

Amusan, L., and S. Oyewole. 2017. The quest for hegemony and the future of African solution to African development problems: Lessons from headways in African security sector. *Journal of Asian and African Studies* 52 (1): 21–33.

Amusan, L., and S. Oyewole. 2023. Precision agriculture and the prospects of space strategy for food security in Africa. *African Journal of Science, Technology, Innovation and Development* 15 (3): 325–336.

Argoun, M.B. 2012. Recent design and utilization trends of small satellites in developing countries. *Acta Astronautica* 71: 119–128.

Bayles, M.D., ed. 1968. *Contemporary Utilitarianism*. New York: Anchor Books.

Bentham, J. 2000 [1781]. *An Introduction to the Principles of Morals and Legislation*. Ontario: Batoche Books Limited.

Bhaskaranarayana, A., C. Varadarajan, and V.S. Hegde. 2009. Space-based societal applications—relevance in developing countries. *Acta Astronautica* 65: 1479–1486.

Bochinger, S. 2016. Space Exploration in the Space Economy. Paper presented at Symposium on the Role of Industry in Space Exploration, United Nations, Vienna, February 15.

Boroffice, R.A. 2008. The Nigerian space programme: An update. *African Skies* 12: 40–45.

Boroffice, R.A., and G.I. Agbaje. 2006. Deploying space research to energy development in Nigeria. *Nigerian Journal of Space Research* 2: 101–113.

Bruni, L., and P.L. Porta. 2007. *Handbook on the Economics of Happiness*. Massachusetts: Edward Elgar Publishing Inc.

Burger, E., and G. Bordacchini, eds. 2019. *Yearbook on Space Policy 2017: Security in Outer Space: Rising Stakes for Civilian Space Programmes*. Geneva: Springer.

de Klerk, H.M., and G. Buchanan. 2017. Remote sensing training in African conservation. *Remote Sensing in Ecology and Conservation* 3 (1): 7–20.

Detsis, B., and E. Detsis. 2013. The benefits brought by space—general public versus space agencies perspectives. *Acta Astronautica* 88: 129–137.

Dick, S. J. (ed.). 2015. *Historical Studies in the Societal Impact of Spaceflight*. Washington, DC: National Aeronautics and Space Administration (NASA).

Dolman, E.C. 2002. *Astropolitik: Classical Geopolitics in the Space Age*. London: Frank Cass.

Edward, W. 1992. *Utility Theories: Measurements and Applications*. New York: Springer.

Euroconsultant. 2017. Government Spending in Space Programs Reaches $62 Billion in 2016, May 30.

Fashade, O.A. 2008. National and regional capacity building activities for the acquisition of rocket launch capacity and the associated infrastructure—Nigerian strategy. *African Skies* 12: 86–89.
Froehlich, A. 2019. *Embedding Space in African Society: The United Nations Sustainable Development Goals 2030 Supported by Space Applications*. Cham: Springer.
Froehlich, A., and A. Siebrits. 2019. *Space Supporting Africa: A Primary Needs Approach and Africa's Emerging Space Middle Powers*. Cham: Springer.
Froehlich, A., N. Ringas, and J. Wilson. 2020. *Space Supporting Africa: Security, Peace, and Development through Efficient Governance Supported by Space Applications*. Cham: Springer.
Gottschalk, K. 2008. Roles of Africa's institutions in ensuring Africa's active participation in the space enterprise: The case for an African space agency. *African Skies* 12: 26–28.
Gottschalk, K. 2010. South Africa's space program. *Astropolitics* 8 (1): 35–48.
Gottschalk, K., P. Waswa, and S. Oyewole. 2018. The African Space Race. Filling Space, September 28. https://filling-space.com/2018/09/28/the-african-space-race/
Harding, R.C. 2013. *Space Policy in Developing Countries: The Search for Security and Development on the final frontier*. New York: Routledge.
Harding, R.C. 2016. Introduction: Space policy in developing countries. *Space Policy* 37: 1–2.
Harvey, B., H.H.F. Smid, and T. Pirard. 2010. *Emerging space powers: The New space programs of Asia, the Middle East and South America*. New York: Springer.
HDoA (Headquarters Department of the Army). 2005. Space Support to Army Operations. Field Manual No. 3–14, Washington, DC, 18 May 2005.
Hudgins, E. L. (ed.). 2002. *Space: The Free-Market Frontier*. Washington DC.: CATO Institute.
Isoun, T. T., and M. J. Isoun. 2013. *Why Run Before Learning to Walk?: Reflections on High Technology as a strategic tool for development in Nigeria*. BookBuilders.
James, G., J. Akinyede, and S. Halilu. 2014. The Nigerian space program and its economic development model. *New Space* 2: 23–29.
Keith, A. 2014. Emerging space programs spark earth observation growth. *Earth Imaging Journal*.
Klein, J.J. 2006. *Space Warfare: Strategy, Principles and Policy*. London and New York: Routledge.
Koekemoer, J., P. Bakkes, S. Mostert. 1999. Distributed telemetry & telecommand for future sunsat missions. *Acta Astronautica* 44(7–12): 531–541.
Lal, B., E. J. Sylak-Glassman, M. C. Mineiro, N. Gupta, L. M. Pratt, and A. R. Azari. 2015. *Global Trends in Space Volume 1: Background and Overall Findings*. IDA Paper P-5242, vol. 1 (June). Washington, DC: IDA Science & Technology Policy Institute.
Luncedo, N., and K. Gottschalk. 2013. The growth of space science in African countries for Earth observation in the 21st century. *South African Journal of Science* 109 (1–2): 1–5.
Lyons, D. 1965. *Forms and Limitations of Utilitarianism*. Oxford: Oxford University.
Maini, A.K., and V. Agrawal. 2014. *Satellite Technology Principles and Applications*, 3rd ed. West Sussex: Wiley.
Martinez, P. 2009. Building capacity in the basic space sciences in Southern Africa: Experiences and prospects. *Advances in Space Research* 43: 1866–1872.
Martinez, P. 2012a. The African leadership conference on space science and technology for sustainable development. *Space Policy* 28: 33–37.
Martinez, P. 2012b. Is there a need for an African space agency? *Space Policy* 28: 142–145.
Martinez, P. 2016. The development and initial implementation of South Africa's national space policy. *Space Policy* 37: 30–38.
Mill, J. S. 2001 [1863]. *Utilitarianism*. Ontario: Batoche Books Limited.
Mistry, D., and B. Gopalaswamy. 2012. Ballistic missiles and space launcher vehicles in regional powers. *Astropolitics* 10 (2): 126–151.
Mohammed, S.O. 2011. *Nigerian Space Research and Development Agency. Director-General Presentation*. Abuja: NASRDA.
Moltz, J.C. 2014. *Crowded Orbit: Conflict and Cooperation in Space*. New York: Colombia University Press.
Momoh, W. 2008. An overview of Nigerian space activity and space law. *African Skies* 12: 104–108.

Mostert, S., and M. Jacobs. 2008. ARM constellation—establishing a regional remote sensing asset. *Acta Astronautica* 63: 221–227.

Mostert, S., and J. Koekemoer. 1997. The science and engineering payloads and experiments on sunsat. *Acta Astronautica* 41 (4–10): 140–411.

Mostert, S., H. Steyn, H. Burger, and H. Bosman. 2008. Sumbandilasat—an operational technology demonstrator. *Acta Astronautica* 63: 1273–1282.

Mowthorpe, M. 2004. *The Militarization and Weaponization of Space*. Lanham: Lexington Books.

Mulgan, T. 2007. *Understanding Utilitarianism*. Stocksfield: Acumen Publishing Limited.

Munsami, V. 2014. South Africa's national space policy: The dawn of a new space era. *Space Policy, 30* 2014: 1–6.

Munsami, V., and E. O. Offiong. 2017. Should Africa follow the European space governance model for an African space programme? *Space Policy* 30: 1–6.

Ng, Y., and L.S. Ho. 2006. *Happiness and Public Policy: Theory, Case Studies, and Implications*. New York: Palgrave Macmillan.

Offiong, E., and V. Munsami. 2018. Towards a space policy institute for Africa. *Space Policy* 48: 23–29.

Ogunnubi, O., and S. Oyewole. 2020. Africanism, power politics and international relations in Africa. In *Power Politics in Africa: Nigeria and South Africa in Comparative Perspective*, ed. O. Ogunnubi, and S. Oyewole. Newcastle: Cambridge Scholar Publishing.

Oikonomou, I. 2017. 'All u need is space': Popularizing EU space policy. *Space Policy* 30, 1–7.

Ojo, J.S. 2017. Geo-spatial distribution of cloud cover and influence of cloud induced attenuation and noise temperature on satellite signal propagation over Nigeria. *Advances in Space Research* 59: 2611–2622.

Olla, P., ed. 2009. *Space Technologies for the Benefit of Human Society and Earth*. New York: Springer, Berlin Heidelberg.

Oyewole, S. 2017. Space research and development in Africa. *Astropolitics* 15 (2): 185–208.

Oyewole, S. 2020a. The quest for space capabilities and military security in Africa. *South African Journal of International Affairs* 27 (2): 147–172.

Oyewole, S. 2020b. Military capabilities of regional powers in Africa: Nigeria and south Africa in comparative perspective. In *Power Politics in Africa: Nigeria and South Africa in Comparative Perspective*, ed by Olusola Ogunnubi and Samuel Oyewole. Newcastle: Cambridge Scholar Publishing Ltd.

Oyewole, S. 2021. Air power capabilities and applications of African regional powers. Presentation at the (British) Royal Air Force Museum Conference on "New Thinking in Air Power, September 16th".

Oyewole, S. 2022a. How space research can help Africa meet the SDGs. Presentation at a webinar of the University of Johannesburg's Institute for Pan-African Thought and Conversation (IPATC) on "Reaching for the final frontier: The African Space Sector", September 29.

Oyewole, S. 2022b. Air and space strategies in African disaster management. Virtual Lecture of the (British) Royal Air Force (RAF) Museum on August 11.

Oyewole, S. 2022c. Air and space strategy in African Counterinsurgency Operations: Insights from the Nigeria's War against Boko Haram. Presentation at the International Conference on Counter-terrorism, Technology and Development in Africa, jointly organised by the Centre for Military Studies, Stellenbosch University, South Africa and the Africa Research Institute, Óbuda University, Hungary, September 22.

Oyewole, S. 2023a. The African orbital journey in space age: From spectator to participant. *Vanguardia Dossier* 88 (15): 72–75.

Oyewole, S. 2023b. Military space strategies and African realism: Egypt, South Africa and Nigeria. In *Oxford Handbook of Space Security*, ed S. M. Pekkanen and P. J. Blount. Oxford: Oxford University Press.

Pelton, J.N., S. Madry, and S. Camacho-Lara. 2017. *Handbook of Satellite Applications*, 2nd ed. New York: Springer.

Preston, B. et al. 2002. *Space Weapons Earth Wars*. Santa Monica, CA: Rand.

Sadeh, E., ed. 2002. *Space Politics and Policy: An Evolutionary Perspective*. Dordrecht: Kluwer Academic Publishers.

Sadeh, E., ed. 2013. *Space Strategy in the 21st Century*. New York: Routledge.

SANSA. 2014a. *South Africa Remote Sensing*. SANSA

Scarre, G. 1996. *Utilitarianism*. London and New York: Routledge.

Schoonwinkel, A., G.W. Miine, and S. Mostert. 2003. Opportunities in satellite based earth observation, telecommunication and navigation for sustainable development in Southern Africa. *Acta Astronautica* 53: 749–759.

Schultz, B. 2017. *The Happiness Philosophers: The Lives and Works of the Great Utilitarians*. Princeton: Princeton University Press.

Sekhula, P. 2013. The law governing the acquisition and use of earth observation data in South Africa: A need for legal harmonisation. *South African Journal of Geomatics* 2 (2): 139–151.

Sen, A., and B. Williams. 1992. *Utilitarianism and Beyond*. Cambridge: Cambridge University Press.

Shabbir, Z., and A. Sarosh. 2018. Counterspace operations and nascent space power. *Astropolitics* 16 (2): 119–140.

Smart, J.J.C., and B. Williams. 1975. *Utilitarianism: For and Against*. Cambridge: Cambridge University Press.

Smith, K. 2015. Africa as an agent of international relations knowledge. In *Africa and International Relations in the 21st Century*, ed S. Cornelissen, F. Cheru and T. M. Shaw. New York: Palgrave Macmillan.

Space in Africa. 2019a. *African Space Industry Annual Report* (2019 Edition). Lagos.

Space in Africa. 2021a. African space industry revenue to surpass USD 10.24 billion by 2024 despite Covid-19 setback, July 6

Space in Africa. 2021b. A review of 283 upstream and downstream companies in the African space and satellite industry, July 20.

Space in Africa. 2022. African space and satellite industry now valued at USD 19.49 billion, August 17.

Tella, O. 2018. Space as a fulcrum of Nigeria's external relations and regional hegemony. *Space Policy* 46: 46–52.

The Tauri Group. 2011. *The South African Space Industry*. Cape Town: Foundation for Space Development South Africa (FSDSA).

UCS (Union of Concerned Scientists). 2020. Satellite Database. Available at: https://www.ucsusa.org/resources/satellite-database. Accessed on 28 Jan 2020

UN (United Nations). 2002. *United Nations Treaties and Principles on Outer Space*. New York: United Nations Publications.

UNDDA (United Nations Department of Disarmament Affairs). 1991. *South Africa's Nuclear-Tipped Ballistic Missile Capability*. New York: United Nations Publications.

UNOOSA. 2021. Space Economy Initiative, 2020 Outcome Report, January.

Van Riper, A.B. 2004. *Rockets and Missiles: The Life Story of a Technology*. Westport: Greenwood Press.

van Wyk, J.A. 2009. South Africa's space policy and interests: A new dawn or a black hole? *Strategic Review for Southern Africa* 31 (2): 46–73.

van Wyk, J. A. 2014a. Apartheid South Africa's nuclear weapons programme and its impact on Southern Africa. *Austral: Brazilian Journal of Strategy & International Relations* 3(6): 119–140.

Veenhoven, R. 2004. Happiness as an aim in public policy: The greatest happiness principle. In *Positive Psychology in Practice*, ed A. Linley and S. Joseph. Hoboken, N.J.: John Wiley and Sons, Inc.

Weinstein, D. 2007. *Utilitarianism and the New Liberalism*. Cambridge: Cambridge University Press.

Wolter, D. 2006. *Common Security in Outer Space and International Law*. Geneva: United Nations Publications.

Wood, D., and A. Weigel. 2011. Building technological capability within satellite programs in developing countries. *Acta Astronautica* 69: 1110–1122.

Wood, D., and A. Weigel. 2012. Charting the evolution of satellite programs in developing countries: The space technology ladder. *Space Policy* 28: 15–24.

Wood, D., and A. Weigel. 2014. Architectures of small satellite programs in developing countries. *Acta Astronautica* 97 (2014): 109–121.

2 Space Policy, Development, and Security: Conceptual Framework of Analysis

Abstract

This chapter offers relevant insights into debate and clarity on key concepts that underscore the subject of this book, such as space, space policy, development and security. On space, the chapter provides insights on the debates between pure and applied scientists on one hand and humanities and social scientists on the other hand, the resultant emergence of geometric and social spaces, and the perspectives on absolute, relative and relational spaces. This is important to reiterate the multidisciplinary perspectives that are required in this study. Although space (outer space) is generally geometric and specifically astronomical, it is interlocked with social spaces in its policy context. On this basis, this chapter align with utilitarian conception of space policy as the use of space and attendant technologies to promote the socioeconomic development and security of the affected party. Accordingly, the concepts of development and security are interrogated and then summarised as the aspirations of the affected party, that is, the people and government concerned.

2.1 Introduction

This volume is situated at the intersection of multidisciplinary discourse that involves African studies, political and policy studies, international relations, development studies, security studies, science and technology policy, space politics and policy studies, and Third world/Global-South/emerging countries studies. These are disparate disciplines with convergent and divergent perspectives on key concepts under consideration. As a matter of divergent views, African studies are dominantly shaped by students of humanities and social sciences, which are more familiar with social dimension of space—broadly covering political, economic, social and cultural dimensions of public and private realms—than astronomical space and other forms of geometric dimension of space (Oyewole 2020).

S. Oyewole, *Utilitarianism in Outer Space*, Southern Space Studies,
https://doi.org/10.1007/978-3-031-49646-2_2

Equally, economic growth and transformation are critical to the understanding of development in neoliberal scholarship championed by the advanced nations of the Global-North, while independent national capacity is prioritised in post-colonial perspective championed by the emerging countries of the Global-South (Martinussen 1997; Pieterse 2010).

The convergence of multiple disciplines in a research of this nature is premised on the realisation that knowledge is relational; and as such, understanding of a given phenomenon is defined by its distinguished content and context in connection to every other objects and subjects in time and space (Oyewole et al. 2023). Hence, space policy cannot be fully understood outside the general context of physical, natural and social forces that underscored the cosmic science and technologies, as well as the specific context of development and security strategies of its sponsor. Against this background, this chapter examines the key concepts in this book, including space, space policy, development, and security. These concepts lack universally acceptable meanings, and require multidisciplinary approach. The importance of this chapter cannot be underestimated, considering the multidisciplinary audience targeted by the book. For this purpose, this chapter is broadly divided into five sections, including space, space policy, development, security, and conclusion.

2.2 Space

If policy is detached from space policy in the context of space policy, we will be left with the concept of space that deserves a considerable clarification. Space is an important concept that has attracted interdisciplinary scrutiny and interpretation. Aristotle (1928) observed that space take precedence over all other things, because without it nothing else can exist. It is no surprise therefore that there is no universal consensus on what constitutes a space. As "one of those words that frequently elicits modification", Harvey (2004:2) described the challenges of operationalizing the concept of space as follows:

> "The complications perhaps arise more out of the modifications (which all too frequently get omitted in the telling or the writing) rather than out of any inherent complexity of the notion of space itself. When, for example, we write of "material", "metaphorical", "liminal", "personal", "social" or "psychic" space (just to take a few examples) we thereby indicate a considerable diversity of contexts which so inflect matters as to seem to render the meaning of space itself entirely contingent upon the context. Similarly, when we designate its range of applications in terms, such as spaces of fear, of play, of cosmology, of dreams, of anger, of particle physics, of capital, of geopolitical tension, of hope, of memory, or of ecological interaction (again, just to indicate a few of a seemingly infinite range of potential sites of deployment of the term) then we seem to be saying that the arena of application defines something so special about the meaning of space as to render any general consideration of its properties a hopeless task."

The divergent perspectives on the concept of space can be narrowed down to a series of debates between physicists and philosophers (Disalle 2006:1). At the broadest level, the concept of "space" is a source of debate between social sciences and humanities on one hand as well as pure and applied sciences on the other. Humanities and social sciences, following the philosophical understanding of space, situate the concept of space in social and relational context; while pure and applied sciences align with the physical understanding of space, which is mostly considered to be absolute and geometric (Lefebvre 1991). Accordingly, there are social and geometric dimensions to space and the conceptions of the two vary in terms of tangibility and measurability. Harvey (2004), therefore, argued that both social and geometric conceptions of space can fit into the understandings of space in absolute, relative and relational sense.

2.2.1 Perspectives on the Concept of Space

2.2.1.1 Absolute Space

The notion of absolute space is mainly associated with Newton and it is usually represented as a geometrically Euclid, fixed frame for event planning, pre-existing and immoveable grid amenable to cadastral mapping and standardized measurement, and open to calculation and engineering practices (Harvey 2004:2; Disalle 2006:13–20). Socially this is the space of private property and other bounded territorial designations (such as states, administrative units, city plans and urban grids) (Harvey 2004:2).

2.2.1.2 Relative Space

Einstein developed the idea of relative space. Prior to this, Kant has provided ample criticism of absolute space and German mathematician and Physicist Carl Friedrich Gauss (1777–1855) has developed rules of non-Euclidean spherical geometry. The central argument is that all forms of measurement depended upon the frame of reference of the observer. It is, therefore, considered impossible to understand space independent of time under this formulation and this mandates an important shift of language from space and time to space–time or spatiotemporality (Harvey 2004; Disalle 2006).

2.2.1.3 Relational Space

The concept of relational space, which is largely associated to Leibniz, also emerged as a criticism against Newtonian absolute space. The relational view of space holds that there is no such thing as space outside of the processes that define it. Processes do not occur in space but define their own spatial frame (Harvey 2004). Like the relative space, this conception also supported the fusion of time and space in analysis. Hence, an event or a thing at a point in space cannot be understood by appeal to what exists only at that point in time, but through everything else going on around it (although in practice usually within only a certain range of influence). A wide variety of disparate influences swirling over space in the past, present and future, concentrate and congeal at a certain point to define the nature of that point (Harvey 2004). This will later help us to understand the connection between space and policy.

2.2.2 Dimensions in Conception of Space

2.2.2.1 Social Space

Social space can be described as structure of the system that gives meaning to the struggles of individuals (Healy 2007). At the individual level, the life course is a sequence of statuses and roles achieved by particular people, together with their own representations of their biographies. At the macro level, it is a more-or-less institutionalized set of roles (and rules governing transitions between them), together with collective representations of what a well-ordered life ought to look like (Healy 2007:107). It is described by Zieleniec (2007: xii) as the territory of self-actualising techniques, of developmental and faith-based systems of meaning, promised enlightenment and the expansion of the minds' horizons beyond the corporeal or physical world of existence. It is relevant to explain developmental puzzles and predict suitable ways forward for political, economic, social and cultural affairs or domain of a group of people, society, the state, and at the broadest level the international community.

2.2.2.2 Geometric Space

Geometric space involves the interval of time, or distance between lines, objects or locations (Abhyankar and Sathaye 1974). This is applicable to mathematics, geography and physics.

2.2.3 Range of Application

Space in this study is outer space. Hence, the range of application of the concept of space is generally geometric and specifically astronomical as well as relative. In this consideration, space is a place (Lambright 2003). This is a frontier that can be related to land, water/seas, air and cyberspace domains (HDoA 2005; Schwartz 2017; Oyewole and Duyile 2021). Space is simply a region between all astronomical objects in the universe (Encarta Dictionaries 2009). Scientific enquiry in astronomical space advances human knowledge of planet Earth, its atmospheres and satellite, the solar system, the galaxies and the universe (Hawking and Mlodinow 2010; Mo et al. 2010; Seeds and Beckman 2010; Sparke and Gallagher 2007). In the solar system, space or distance is often calculated using the standard astronomical unit (au): the mean distance from the centre of the Sun to the centre of the Earth, which denotes 149,597,870.691 km (km) plus or minus (±) 30 m (m), or 92,955.807.267 international miles ± 100 ft. (Vanderbei and Belikov 2007:91). Also, light-year—the distance light travel for one year—is a common unit used by astronomers to describe distances in the universe. Light travels at the pace of 3×10^5 km per second (km/s), making a light-year 9.46×10^{12} km (9,460,000,000,000 km) (Fraknoi et al. 2017:16).

Space is also the infinite region that lies beyond the Earth's atmosphere, and all that it contains (Encarta Dictionaries 2009). This definition loosely includes every astronomical object except planet Earth as space. Zieleniec (2007:xi) described this space as "somewhat outside this world, somewhere up there." Most of the United Nations' (UN) resolutions, principles, and treaties on space use the word to cover all celestial bodies (UN 2002). Many national space policy papers and legislations follow this pattern. However, the Earth's atmosphere is made up of multiple layers, which start from troposphere, between 8 and 17 km altitude (height above the ground or sea level); stratosphere, 17 to 50 km; mesosphere, 50 km to about 80 km; thermosphere, 80–500 km; and end with exosphere, 500–1000 km altitude (Moore 2002:40). Thus, there are different postulations on where space starts from the Earth point of view. Using different criteria, the boundary of space from the Earth have been postulated to starts from 80, 100, 118, 122, 129–150, and 400–800 km altitudes (Lal and Nightingale 2014; Oyewole 2017). Amidst these, the 100 km altitude, postulated by Theodore Von Karman as the boundary between aeronaut and astronaut and taken as the starts of space from the Earth, and named after him as Karman line, is widely in use and accepted by the International Astronautical Federation (IAF) (Macdonald 2014).

2.3 Space Policy

Although the understanding of space adopted in this study is relative to human knowledge of the Earth and its location in the universe, space policy mean space in relation to policy. Policy falls into the class of social space, while outer space fit into geometric concept of space. Accordingly, the knowledge of the concept of policy is important to the proper understanding space policy. Cochran and Malone (2014:3) defined policies as purposive courses of action devised in response to a perceived problem. This is a general definition that gives no hint on the likely actor and problem, and overlooks the possibility of having a response designed based on perceived opportunity instead of problem. In the light of this, space policy can be described as a purposive course of action devised in response to actual or perceived scientific, technological, socio-economic, political, and military challenges or opportunities in or from outer space. In this consideration, the action can be to explore or exploit the outer space, the space actor can be state or non-state actors, the context can also be public or private, and at individual, domestic and international levels.

In a limited context, space policy can be considered as a form of public (national) policy, such as the health, economic, defence and foreign policies. Cochran and Malone (2014:3) defined public policy generally as the overall framework within which government actions are undertaken to achieve public goals, and specifically government decisions and actions designed to deal with a matter of public concern. In the context of outer space, Tronchetti (2013: IX) observed that the term policy refers to the official approach of a state towards the exploration and use of outer space. Space policies include both the making of space policy through the legislative process and the execution of that policy by civilian, military bodies, and regulatory agencies (Tronchetti 2013: ix). This definition helps to identify major policy actions (policy making and executions) that are required in space and the actors in each case.

Sadeh (2002) shares Tronchetti's (2013: ix) perspective, with more details, on the concept of space policy. Sadeh (2002: xiv) describes space policy as both the process (space politics) of policy formation and policy change over time, and the courses of action taken to achieve political and technological determined outcomes (space policy). In the light of this, Sadeh (2002) observed that space politics involves the process by which historic conditions, rationales for space, and advocacy coalitions interact with and impact agenda-setting; actors and institutions (Presidents, Congress, and the space bureaucracy) interact with and impact public policy formulation and implementation; and how policy outcomes bring about policy change (emergence of privatization and commercialization). Equally, space policy deals with the outcomes that include such areas as the environment, law, commerce, international cooperation, and national security (Sadeh 2002: xiv). This understanding further sharpens the place of space–time and relational meaning of outer space policy. However, this conception left someone to wonder why space policy broadly equates input and output at one hand, and on the other hand output alone also equates to space policy.

Tronchetti's (2013: ix) argued that a "space policy" describes a nation's strategy regarding its civilian space programme and the military and commercial utilization of outer space. As such, Klein (2006:137) observed that policy commonly provides guidance for governmental departments and agencies, while delineating resources to make political goals a reality. In the light of these, some scholars have tried to differentiate between space policy and space strategy (Klein 2006; Sadeh 2013). Although policy includes elements of strategy, the two concepts are considerably different (Akinrinde et al. 2023; Cochran and Malone 2014). Policy is the official governmental positions, either in the form of documents or speeches, which provide the necessary guidance to reconcile competing requirements in order to balance one's "ends" with one's limited "means" (Klein 2006:137). Strategy involves one's ends and means, and how they are managed. It is the organisation and utilisation of relevant resources in order to achieve the stated objectives of a given actor (Akinrinde et al. 2023). Sadeh (2013:1) argued that the purpose of space strategy is to coordinate, integrate, and prioritize space activities across security, commercial, and civil sectors. Without strategy, space activities continue to provide value, but it becomes difficult to identify and execute long term programs and projects and to optimize the use of space for security, economic, civil, and environmental ends (Sadeh 2013:1). Nevertheless, the concepts of space policy and strategy are too overlapping that the difference between the two is not well pronounce and the idea and attempts to distinguish them are less significant in many considerations.

This study generally aligns with utilitarian conception of space policy. In this manner, Harding (2013:5) concisely defines national space policy as the conscious and specific use of space and its attendant technologies to promote the security and socioeconomic (development) interests of the state. This definition identified four major concepts that need to be understood in any conception of space policy, and these are space, space technology, security and socio-economic interests of the sponsor (state). The concept of space was earlier considered in this chapter. Space technology, however, broadly include satellite system (including orbital and ground components), rocket and orbital vehicle, astronomical observatories, and all related facilities and infrastructures. Harding (2013) used the concept of socioeconomic interests of the state interchangeably with development, which will be examined in the next section before security.

2.4 Development

The concept of development has multiple meanings across time and space. Hence, it is a subject of academic and public debate. Hettne (2005:27) observed that there is always this tendency for official, politically-recognised and generally acceptable understanding to emerge among theoreticians and practitioners to undermine other perspectives, hence making control over discourses a dimension of power. Accordingly, a distinction is made between *mainstream* and *counterpoint* on the concept of development, the former referring to the predominant view, focusing on modernisation and economic growth from either a state or a market perspective. The latter refers to certain oppositional, sometimes "anti-modern" ideas from "civil society", typically arguing for inherently superiority of small-scale, decentralised, ecologically sound, community-centred, human and stable models of societal development (Hettne 2005:27). Studies on means and ends of development have shed more light to this dichotomy and their historical trajectories (Martinussen 1997; Pieterse 2010). In this manner, Pieterse (2010) provides a historical review of the understanding of development between 1800 and 2000s (see Table 2.1).

In relation to the meaning of development, classical political economy in the 1800s largely focus on how to address shortcomings and maladies of progress, such as population growth, social inequalities, social dislocation and wealth distribution, from where Thomas Robert Maltus (1766–1834) and Karl Marx (1818–1883) among others emerged prominent (Pieterse 2010). As such, Marx's conception of development entails a proletarian revolution that will replace capitalism with socialism, and with the final aim of achieving communism. Similarly, many social scientists are increasingly responding as well as calling for more responses to political, economic, legal and cultural questions, including opportunities and challenges, which are associated with the advancement in pure and applied sciences as well as engineering that underscored space exploration and exploitation with all their various space problematisation perspectives and attendant choice of actions contained in recommendations (Dick 2015; Pass 2006; Wolter 2006).

Table 2.1 Meanings of development over time

Period	Perspectives	Meanings of development
1800s	Classical political economy	Remedy for progress, catching up
1870>	Latecomers	Industrialization, catching-up
1850>	Colonial economics	Resource management, trusteeship
1940>	Development economics	Economic growth – industrialization
1950>	Modernization theory	Growth, political and social modernization
1960>	Dependency theory	National auto-centric accumulation
1970>	Alternative development	Human flourishing
1980>	Human development	Enlargement of people's choices and capacity
1980>	Neoliberalism	Economic growth—structural reform, deregulation, liberalization, privatization
1990>	Post-development	Authoritarian engineering
2000>	Millennium development goals	Structural reforms, global good
2015>	Sustainable development goals	Common good across generations

Source Pieterse's (2010:7) revised and updated by the author

With the dawn of industrial revolution in Western Europe, industrialisation and catching up became the dominant ends of development, particularly in the rest of the continent, and most especially from 1870s (Pieterse 2010). This conception of development is equally relevant to developing countries of the Global-South in their bids to catch up with developed countries of the Global-North, in this case, the advanced space players. For example, Harding (2013) observed that developing countries, such as China, India and Brazil are first-tier space players are playing catch up with the developed countries and advanced space players, such as the USA, Russia and Western Europe. Trusteeship and access to colonial resources were also considered as development agenda from 1850s (Pieterse 2010). This is related to the growing advocacy for human colony or settlement and resources exploitation in space and celestial bodies (Schwartz 2017). In fact, there are space advocates that believed that space colonialism is an unavoidable stage of development that is underscore by social, political and economic trends. In the same way that colonialism was considered exportation of civilisation by the metropolis to the colonies, some advocates also observe the need to take civilisation to other worlds (Schwartz 2017).

Since the 1940s, the concept of development has emerged prominent out of what used to be conceived as progress and evolution (Martinussen 1997:34). Amidst these, economic growth and industrialisation were popularly considered as ends of development in 1940s (Pieterse 2010). This conception of development remains popular, despite its well-known inadequacy, most especially in the developing countries. In relation to this conception of development, however, the potentials of space exploration and exploitation to inspire economic growth have been recognised (Harding 2013).

From the 1950s, modernisation thesis championed by Western scholars encouraged developing countries to follow Western footsteps and adopt their technologies, socio-economic and political systems to achieve development (Martinussen 1997; Pieterse 2010). In response, dependency or post-colonial thesis emerged prominently in the 1960s to encourage developing countries to prioritise national self-reliance and independent capacity in technological and socioeconomic development (Akinrinde and Oyewole 2021). The modernisation conception of development is relevant in the attempt to understand the growing demand for space products and technological transfer from the advanced space players of the Global-North to the Global-South, while the post-colonial perspective explains the drive for independent national capacity in space technologies by several developing countries (Argoun 2012; Harding 2013; Wood and Weigel 2011, 2012).

Starting from 1970s, the foregoing state-centric conceptions of development have been criticised by advocates of alternative development (Pieterse 2010). Amidst these, community development was conceived with ends of human flourishing. Philips and Pittman (2009) defined community development as "a *process*: developing and enhancing the ability to act collectively, and an *outcome*: (1) taking collective action and (2) the result of that action for improvement in a community (a place or interest group) in any or all realms: physical, environmental, cultural, social, political, economic, etc." Equally, human development has emerged since 1980s with focus on enlargement of people's choices (UNDP 1990:10; Pieterse 2010; Martinussen 1997). Space and attendant technologies are relevant in this consideration for job creation and poverty alleviation, urban planning, precision agriculture and food production, as well as management of water, transportation, health, and environment, among other things that are strategic to community and human developmental interests (Dick 2015; Olla 2009; Froehlich 2018; Oyewole 2022).

From 1980s, neoliberalism has become a popular perspective with emphasis on economic growth based on structural reform, deregulation, liberalisation and privatisation which are aimed to roll back government and reduce market-distorting interventions and in effect lead or inspire 'development' (Pieterse 2010). This is important to liberalisation and the growing involvement of private firms in space exploration and exploitation (Hudgins 2002; Pelton 1992). Nevertheless, neoliberalism is also considered as an anti-development perspective, not in terms of objectives but in terms of means. Post-development thinking has therefore puts forth an anti-development position, which is more radical in orientation and applies not merely to the means (the state is accused of authoritarian engineering) but also to the goals (economic growth is repudiated) and the results (which are deemed a failure or disaster for the majority of the population) (Pieterse 2010:7). This perspective is salient in the Global-South, where market turn out to be a major socioeconomic disaster, especially in the 1980s and 1990s (Amusan and Oyewole 2012; Stiglitz 2002). This perspective is equally relevant to understand the concerns, scepticisms and criticisms amidst the optimisms and prospects of private corporations, their interests and activities in space.

The concept of development has taken new momentum since the United Nations (UN) adopted vision 2015, Millennium Development Goals (MDGs) in 2000 (Pieterse 2010). This programme was conceptualised to entails eight different goals, which include to (1) eradicate extreme hunger and poverty, (2) achieve universal primary education, (3) promote gender equality and empower women, (4) reduce child mortality, (5) improve maternal health, (6) combat HIV/aids, malaria and other diseases, (7) ensure environmental sustainability and (8) develop a global partnership for development (UN 2001). The MDG was replaced with a broader vision 2030, Sustainable Development Goals (SDGs) in 2015 (UN 2015). This entails 17 goals, including the following.

To (1) end poverty in all its forms everywhere; (2) End hunger, achieve food security and improved nutrition and promote sustainable agriculture; (3) Ensure healthy lives and promote well-being for all at all ages; (4) Ensure inclusive and equitable quality education and promote lifelong learning opportunities for all; (5) Achieve gender equality and empower all women and girls; (6) Ensure availability and sustainable management of water and sanitation for all; (7) Ensure access to affordable, reliable, sustainable and modern energy for all; (8) Promote sustained, inclusive and sustainable economic growth, full and productive employment and decent work for all; (9) Build resilient infrastructure, promote inclusive and sustainable industrialization and foster innovation; (10) Reduce inequality within and among countries; (11) Make cities and human settlements inclusive, safe, resilient and sustainable; (12) Ensure sustainable consumption and production patterns; (13) Take urgent action to combat climate change and its impacts; (14) Conserve and sustainably use the oceans, seas and marine resources for sustainable development; (15) Protect, restore and promote sustainable use of terrestrial ecosystems, sustainably manage forests, combat desertification, and halt and reverse land degradation and halt biodiversity loss; (16) Promote peaceful and inclusive societies for sustainable development, provide access to justice for all and build effective, accountable and inclusive institutions at all levels; (17) Strengthen the means of implementation and revitalize the global partnership for sustainable development (UN 2015).

Many scholars have argued and shown that space technologies are important to advance MDGs and SDGs (Dick 2015; Froehlich 2018, 2019; Olla 2009). They provide essential services that support poverty alleviation, equality, justice, peace, innovative economy, inclusive institution, food security, human resilience, quality health and education systems, and promote sustainable environment, water, and energy management, as well as international cooperation. Although Africa is a late comer in space, some of these developmental benefits of space and attendant technologies have become noticeable (Froehlich 2019; Froehlich et al. 2020; Oyewole 2022, 2023).

2.5 Security

The concept of security has generated tense policy debates and controversies across time and space. Dillon (1996) observed that the genealogy of security primarily deals with how order of fear forms a people and its implication for what is political. In view of this, security is a principal device for constituting political order and for confining political imagination within the laws of necessity of the specific rationalities thrown-up by their equally manifold discourses of danger (Dillon 1996:16). In this consideration, evolution theory is of the view that tribal societies developed centralised source of authority with army of men over a given piece of land out of fear of losing the property, owned by individual members or the group collectively, to their neighbours (Fukuyama 2012). At the same time, the state, in social contract theory, is thought to have emerged with monopoly of violence to guarantee security against the perpetual fear of impunity that characterised the state of nature (Hobbes 1651).

Security studies is generally dominated by students of International Relations that are known as realists. Accordingly, the concept of security traditionally means national security. Louw (1978:10) defined national security as the condition of freedom from external physical threats which a nation-state enjoys. Huntington (1957:1), however, offers a broader and policy-driven definition of national security. He observed that national security policy can be considered in three forms: military, internal and situational security policies. Military security policy is the programme of activities designed to minimise or neutralise efforts to weaken or destroy the nation by armed forces operating from outside its institutional and territorial confines (Huntington 1957:1). Internal security policy also deals with threat of subversion, that is, the efforts to weaken or destroy the state by forces operating from within its territorial and institutional confines (Huntington 1957:1). Situational security is mostly concerned with the threat of erosion to the relative power of the state, resulting from long time changes in social, economic, demographic, and political conditions (Huntington 1957:1).

The foregoing perspective on security can be criticised for nationalistic, state-centric reductionism, with most focus on territoriality and regime as principal objects to be protected, at the expense of humanity. On many occasions, this rationale has been used to justify impunity of the state against defenceless people. These among others have contributed to the decline of realism and its conception of security (Booth 2011). Hence, the concept of human security has emerged to prominence with emphasis on the fundamental freedom of the people to live a good life (UNDP 1994). It is important to note that justice, coordination, efficiency, and stability demand that a social arrangement is based on the principle of liberty (Rawls 1999). Human security broadly covers issues of economic, food, health, environmental, community, personal, and political security (UNDP 1994: 24–25). These dimensions of human security are briefly considered below:

- Economic security: It requires an assured basic income-usually from productive and remunerative work, or in the last resort from some publicly financed safety net (UNDP 1994:25). In this consideration, unemployment, underemployment, job insecurity, inflation, high dependency ratio and living below poverty line are threats to human security.

- Food Security: It means that all people at all times have both physical and economic access to basic food (UNDP 1994:25). Food security can be considered at national, community and household levels. At national level, food security implies the condition whereby the country is able to manufacture, import, retain and sustain food needed to support its population with minimum per capita nutritional standards (du Toit 2011:3). At the level of community, food security means the condition whereby the residents in a community can obtain safe, culturally accepted, nutritionally adequate diets through a sustainable system that maximises community self-reliance (du Toit 2011:3). At household level, it is the availability of food in one's home which one has access to. In this consideration, a household is regarded as food secure when the members of the family do not live in hunger or fear of starvation (du Toit 2011:3). Absence of food security will amount to starvation and famine. Amartya Sen (1981:40–41) described starvation as lowness of the typical level of food consumption and declining trend of food consumption, and famine as sudden collapse of the level of food consumption. Food security can be affected by major global concerns, such as environmentally sustainable development, water resources, world trade, human security and rights, climate change and the current pattern of globalization, which penetrate and affect the concerns at the local and national levels (Shaw 2007:384). In general, food security entails availability, accessibility, and affordability of food, which are complex issues that encompass a wide range of interrelated economic, social and political factors that are both internal and external to a nation (Clover 2003; Amusan and Oyewole 2023).
- Health security: Freedom from diseases and the general condition or wellness of body, mind and soles are also essential to human security. Accordingly, poor nutrition, environmental degradation, including water and air pollution, as well as the rise and spread of infectious and parasitic diseases are considered as threats and concerns to human security.
- Environmental security: This focuses on interdependent relations between humanity and the totality of its ecosystem or every other thing in the surrounding, which include air, waters, land, and space as well as the resources therein. In this consideration, environmental degradation is a major threat at the local level, while climate change is a global threat.
- Community security: Most people derive security from their membership of a group, such as a family, community, organization, racial or ethnic group that can provide a cultural identity and a reassuring set of values (UNDP 1994:31). Hence, security of this group is important in human security, although this perspective is also concern about using this platform as a driver of oppression, hatred and terror of members and non-members. Community security is supported by the International Covenant on Economic, Social and Cultural Rights (ICESCR) of 1966, the International Covenant on Civil and Political

Rights (ICCPR) of 1966, and the United Nations Declaration on the Rights of Indigenous Peoples (UNDRIP) of 2007 among others.

- Personal security: This is arguably the most important dimension of human security. It focuses on freedom of individuals from threats from the state (physical torture), threats from other states (war), threats from other groups of people (ethnic tension), threats from individuals or gangs, against other individuals or gangs (like crime and street violence), threats directed against women (such as rape and domestic violence), threats directed at children based on their vulnerability and dependence (child abuse), and threats to self (such as suicide and drug use) (UNDP 1994:30).
- Political security: The emphasis here is on the level of respect accorded to human rights, rule of law and public participation in a given political system under which individuals live. Accordingly, the principle of self-determination has been recognised as a right by extant international conventions and declarations, as evident with the Chatter of the United Nations (UN) of 1945, the Universal Declaration of Human Rights of 1948, the ICESCR of 1966, the ICCPR of 1966, and the UNDRIP of 2007 among others. Article 1(1) of the ICESCR described self-determination as "the rights of all peoples to freely determine their political status and freely pursue their economic, social and cultural development."

Against this background, the concept of human security is usually seen as an attempt to shift security attention from its traditionally state-centred focus to people-centred. However, the possibility of a good life outside a state remains debatable, given the fear of anarchy, which is associated with freedom without social obligation or constrain (Nozick 1974). This sounds like a caution to any attempt to present human security as a mutually exclusive and irreconcilable alternative to state security. Accordingly, the concept of security must accommodate both state and human security concerns. Imobighe (2003) thus argued that security has recently come to be understood as freedom from, or elimination of, threat not only to the physical existence of the state, but also to its ability to self -protection and development, and the enhancement of the general wellbeing of all the people. Beyond social constraint on liberty in form of obligation, justice and stability require a state to guarantee equal scheme of liberty for all (Rawls 1999). Therefore, security broadly requires political, social, environmental, economic, military, and cultural systems that will together give people the building blocks of survival, livelihood and dignity (UNCHS 2003).

2.6 Space, Security and Development

The connectivity of space, security and development is important. Amidst these, the conception of "space security", "security in space", "space in security" or "security dimension of space exploration and exploitation" among others has increasingly follow the same trend along with conception of security as identified above. Accordingly, "space security" was observed by Sheehan (2015:7–8) with the following submission:

> Traditionally, it has been associated with the military security of states, and this is still the predominant understanding of the term. It certainly remains the case that a fundamental aspect of "space security" is the contribution that satellites make to the military security of states and the maintenance of international stability and balances of power, as well as the military threats to satellites and their ability to support international security. However, there has been a widening of the understanding of space security in recent years so that, while the military dimension remains fundamental, other crucial issues have been brought within the scope of the concept… Changes in the understanding of space security mean that it now similarly embraces a broad understanding of the requirement of freedom of access to and use of space by all states who wish to use it for the socioeconomic benefit of their populations.

Socioeconomic benefits of space and its attendant technology covers poverty alleviation, job creation, urban planning, precision agriculture and food security, as well as water, health, environmental, and transportation management (Froehlich 2018; Olla 2009; Oyewole 2017). In this consideration, the relevance of space to military capabilities, law enforcement and disaster management can now be considered together with the issue of food security, water and health management, environmental sustainability, among other subjects of concerns to both state and human security, in the analysis of space policy and strategy in the general framework of security policy and strategy. Hence, the line between development and security has become increasingly blurred.

There are growing academic and policy efforts to bridge the gap between development and security priorities (Stern and Öjendal 2010; Hettne 2010; Buur et al. 2007; Reid-Henry 2011). This is not totally a new development. Extant literatures published in inter-and-post World Wars years have shown that the seed of peace and security cannot germinate without development and economic progress. However, this position and its advocacy has taken new momentum since the end of the Cold War, especially since the 9/11, and the attendant securitisation of failed and collapse state. This informed the presentation of development as an essential tool of security policy. Accordingly, there is a believe that the development discourse and industry is increasingly co-opting security practices in order to strengthen poverty reduction efforts (Denney 2011: 278). Equally, the UNDP considers security as a dimension of human development. These among other factors informed the academic discussion on development-security nexus.

Beyond human security, state-centric security has also co-opted development into efforts to build military and law enforcement capabilities. Tellis et al. (2000) observed that military capabilities cannot be adequately assess without paying attention to the quantity and quality of strategic resources, including human and material resources, that are made available to the military by the society. In essence, the level of education, healthiness, and general standard of living of the population will influence the level of their productivity and its percentage that can be channelled to build the military capabilities. Equally, military spending and industrial development have for a long time rely on the assumption that they contribute to development as justification for their continuous relevance. Besides, human development and security indicators have become indispensable factors to assess the tendency for stability and instability in a political system in recent times (World Bank 2011). Hence, they are important to the law enforcement capability of a state.

2.7 Conclusion

This chapter explores the meanings of key concepts of this study (space, space policy, development and security) using multidisciplinary and multidimensional approaches. In summary, the chapter generally align with geometric dimension of space as the context for astronomical space (outer space), which is admitted in relations to policy and other aspects of social space in the attempt to understand the content and context of space policy. In addition, the changing understanding of the ends and means of development across time and space are interrogated, without losing sight of the dichotomy between the state and the people in conceptions of development and security.

Space policy broadly involves the course of action pursued in space and the politics that produced it. Equally, space politics generally covers the interactions of different factors (opportunities and challenges), their conceptions and assessments by different actors (state and non-state ones) based on their various interests that inspired, conditioned or shaped space programme. Amidst these, what become the policy objectives of a state in space is important vis-à-vis the political process that produce it and associated strategic rationales. Accordingly, the unique conditions of a state and attendant conception of national space policy cannot be ignored, even amidst the appeals of a universalist position that space activities are globally driven by common utilitarian goals, that is, to advance, promote or protect the development and security aspirations of the concerned party. Consequently, it is important to understand the national space policy interests of a country in the context of the unique development and security challenges and opportunities that necessitate them.

References

Abhyankar, S.S., and A.M. Sathaye. 1974. *Geometric theory of algebraic space curves*. New York: Springer-Verlag.

Akinrinde, O.O., and S. Oyewole. 2021. Neo-colonialism and developmental challenges of post-colonial Africa. *Jurnal Pertahanan* 7 (3): 398–410.

Akinrinde, O., S. Oyewole, and B. Omitola. 2023. Introduction to Strategic Studies. In *Introduction to political science: A reading text on politics and governance*, ed. J.S. Omotola, A. Olaniyan, M.B. Aleyomi, and S. Oyewole, 373–384. Oye-Ekiti: Federal University Oye-Ekiti.

Amusan, L., and S. Oyewole. 2012. Global democratisation and capitalism: Discovering the third world states in the era of limited state and unlimited quest. *Canadian Journal of Social Science* 8 (5): 50–57.

Amusan, L., and S. Oyewole. 2023. Precision agriculture and the prospects of space strategy for food security in Africa. *African Journal of Science, Technology, Innovation and Development* 15 (3): 325–336.

Argoun, M.B. 2012. Recent design and utilization trends of small satellites in developing countries. *Acta Astronautica* 71: 119–128.

Aristotle. 1928. *The works of Aristotle translated into English by W.D. Ross.*

Booth, K. 2011. *Realism and world politics.* London and New York: Routledge.

Buur, L., S. Jensen, and F. Stepputat. 2007. The security-development nexus. In *The security-development nexus: Expressions of sovereignty and securitization in Southern Africa.*, ed. L. Buur, S. Jensen and F. Stepputat. Uppsala: Nordiska Afrikainstitutet.

Clover, J. 2003. Food security in Sub-Saharan Africa. *African Security Review* 12 (1): 5–15.

Cochran, C.L., and E.F. Malone. 2014. *Public policy: Perspectives and choices*, 5th ed. Boulder, CO: Lynne Rienner.

Denney, L. 2011. Reducing poverty with teargas and batons: The security-development nexus in Sierra Leone. *African Affairs* 110 (439): 275–294.

Dick, S. J. (ed.). 2015. *Historical studies in the societal impact of spaceflight.* Washington, DC: National Aeronautics and Space Administration (NASA).

Dillon, M. 1996. *Politics of security: Towards a political philosophy of continental thought.* London and New York: Routledge.

Disalle, R. 2006. *Understanding space-time: The philosophical development of physics from Newton to Einstein.* Cambridge: Cambridge University Press.

du Toit, D. C. 2011. *Food security.* Republic of South Africa, Department of Agriculture, Forestry and Fisheries, Directorate of Economic Services, Production Economics unit.

Encarta Dictionaries. 2009. Microsoft® Encarta® 2009. ©1993–2008 Microsoft Corporation.

Fraknoi, A., D. Morrison, and S.C. Wolff. 2017. *Astronomy.* Houston: Rice University.

Froehlich, A. 2018. *Post 2030-agenda and the role of space: The UN 2030 goals and their further evolution beyond 2030 for sustainable development.* Cham: Springer.

Froehlich, A. 2019. *Embedding space in African society: The united nations sustainable development goals 2030 supported by space applications.* Cham: Springer.

Froehlich, A., N. Ringas, and J. Wilson. 2020. *Space supporting Africa: Security, peace, and development through efficient governance supported by space applications.* Cham: Springer.

Fukuyama, F. 2012. *The origin of political order: From prehuman times to the French revolution.* London: Profile Books Ltd.

Harding, R.C. 2013. *Space policy in developing countries: The search for security and development on the final frontier.* New York: Routledge.

Harvey, D. 2004. Space as a key word. Paper for Marx and Philosophy Conference, Institute of Education, London, 29 May.

Hawking, S., and L. Mlodinow. 2010. *The grand design.* New York: Bantam Books.

HDoA (Headquarters Department of the Army). 2005. Space support to army operations. Field Manual No. 3–14, Washington, DC, 18 May 2005.

Healy, K. 2007. Sociology. In *A companion to contemporary philosophy*, vol. 2, 2nd edn, ed. R. E. Goodin, P. Pettit, and T. Pogge. Malden: Blackwell Publishing.

Hettne, B. 2010. Development and security: Origins and future. *Security Dialogue* 41 (1): 31–52.

Hettne, B. 2005. Discourses on development. In *Palgrave advances in development studies*, ed. Jeffrey Haynes. New York: Palgrave Macmillan.

Hobbes, T. 1651. *Leviathan.* London.

Hudgins, E. L. (ed.). 2002. *Space: The free-market frontier.* Washington DC.: CATO Institute.

Huntington, S.P. 1957. *The soldier and the state: The theory and politics of civil military relations.* Massachusetts: Harvard University Press.

Imobighe, T. A. 2003. *Nigeria's defence and national security linkages: A framework of analysis.* Ibadan: Heinemann Educational Books (Nigeria) Plc.

Klein, J.J. 2006. *Space warfare: Strategy, principles and policy.* London and New York: Routledge.

Lal, B., and E. Nightingale. 2014. "Where is Space? And Why Does That Matter?" Space Traffic Management Conference, November 5.

Lambright, W.H., ed. 2003. *Space policy in the twenty-first century.* Baltimore: The John Hopkins University Press.

Lefebvre, H. 1991. *The production of space.* Translated by Donald Nicholson Smith. Oxford: Blackwell

Louw, M. H. H. 1978. Introduction to the national security concept. In *National security: A modern approach*, ed. M. H. H. Louw. Pretoria: University of Pretoria.

Macdonald, M. 2014. A system-level view of space project. In *The international handbook of space technology*, ed. M. Macdonald and V. Badescu. London: Springer Praxis Books.

Martinussen, J. 1997. *Society, state & market: A guide to competing theories of development.* London: Zed Books Ltd.

Mo, H., F.V.D. Bosch, and S. White. 2010. *Galaxy formation and evolution.* Cambridge: Cambridge University Press.

Moore, P. 2002. *Philip's astronomy encyclopaedia: A comprehensive and authoritative A-Z guide to the universe.* London: Philip's.

Nozick, R. 1974. *Anarchy, state, and Utopia.* Oxford: Blackwell Publishers Ltd.

Olla, P., ed. 2009. *Space technologies for the benefit of human society and earth.* New York: Springer, Berlin Heidelberg.

Oyewole, S. 2017. Space research and development in Africa. *Astropolitics* 15 (2): 185–208.

Oyewole, S. 2020. The quest for space capabilities and military security in Africa. *South African Journal of International Affairs* 27 (2): 147–172.

Oyewole, S. 2023. The African orbital journey in space age: From spectator to participant. *Vanguardia Dossier* 88 (15): 72–75.

Oyewole, S., J.S. Omotola, and L. Amusan. 2023. Political science and other fields of studies. In *Introduction to political science: A reading text on politics and governance*, ed. J.S. Omotola, A. Olaniyan, M.B. Aleyomi, and S. Oyewole, 9–19. Oye-Ekiti: Federal University Oye-Ekiti.

Oyewole, S., and A. Duyile. 2021. The enduring relevance of naval power: Reflections on defence, security and power projection of sea frontier. *Austral: Brazilian Journal of Strategy and International Relations* 10(20): 118–133.

Oyewole, S. 2022. How space research can help Africa meet the SDGs. Presentation at a webinar of the University of Johannesburg's Institute for Pan-African Thought and Conversation (IPATC) on "Reaching for the final frontier: The African Space Sector", September 29.

Pass, J. 2006. Astrosociology as the missing perspective. *Astropolitics* 4 (1): 85–99.

Pelton, J. N. 1992. Organizing large space activities: Why the private sector model usually wins. *Space Policy*, 233–244.

Philips, R., and R. H. Pittman. 2009. A framework for community and economic development. In *An introduction to community development*, ed. R. Philips and R. H. Pittman. London and New York: Routledge.

Pieterse, J.N. 2010. *Development theory: Deconstructions/reconstructions*, 2nd ed. Los Angeles: Sage.

Rawls, J. 1999. *A theory of justice.* Revised. Massachusetts: Harvard University Press.

Reid-Henry, S. 2011. Spaces of security and development: An alternative mapping of the security–development nexus. *Security Dialogue* 42 (1): 97–104.

Sadeh, E., ed. 2013. *Space strategy in the 21st century.* New York: Routledge.

Sadeh, E. 2002. Introduction-space politics and policy: An evolutionary perspective. In *Space politics and policy: An evolutionary perspective.* Dordrecht: Kluwer Academic Publishers.

Schwartz, J.S.J. 2017. Myth-free space advocacy part II: The myth of the space frontier. *Astropolitics* 15 (2): 167–184.

Seeds, M.A., and D.E. Beckman. 2010. *Astronomy: Solar system and beyond*, 6th ed. Belmont: Brooks/Cole.

Sen, A. 1981. *Poverty and famines: An essay on entitlement and deprivation.* Oxford: Clarendon Press.

Shaw, D.J. 2007. *World food security: A history since 1945*. New York, NY: Palgrave Macmillan.
Sheehan, M. 2015. Defining space security. In *Handbook of space security*, ed. K.-U. Schrogl et al. New York: New York.
Sparke, L.S., and J.S. Gallagher. 2007. *Galaxies in the universe: An introduction*, 2nd ed. Cambridge: Cambridge University Press.
Stern, M., and J. Öjendal. 2010. Mapping the security-development nexus: Conflict, complexity, cacophony, convergence? *Security Dialogue* 41 (1): 5–30.
Stiglitz, J. 2002. *Globalization and its discontents*. New York: W.W. Norton & Company.
Tellis, A. et al. 2000. *Measuring national power in the post industrial age: Analyst's handbook*. Santa Monica, CA: RAND.
Tronchetti, F. 2013. *Fundamentals of space law and policy*. New York: Springer.
UN. 2002. *United nations treaties and principles on outer space*. New York: United Nations Publications.
UN. 2001. Road map towards the implementation of the United Nations Millennium Declaration. Report of the Secretary-General A/56/326 for UN General Assembly.
UN. 2015. Transforming our world: The 2030 agenda for sustainable developmen nt. A/RES/70/ 1.
UNCHS. 2003. *Human security now*. New York: United Nations Commission on Human Security (UNCHS).
UNDP. 1990. *Human development report 1990*. New York: Oxford University Press.
UNDP. 1994. *Human development report 1994*. New York: UNDP (United Nations Development Programme).
Vanderbei, R., and R. Belikov. 2007. Measuring the astronomical unit. *Sky Telescope* 113 (1): 91–94.
Wolter, D. 2006. *Common security in outer space and international law*. Geneva: United Nations Publications.
Wood, D., and A. Weigel. 2011. Building technological capability within satellite programs in developing countries. *Acta Astronautica* 69: 1110–1122.
Wood, D., and A. Weigel. 2012. Charting the evolution of satellite programs in developing countries: The space technology ladder. *Space Policy* 28: 15–24.
World Bank. 2011. *Conflict, security, and development*. Washington, D.C: The World Bank.
Zieleniec, A. 2007. *Space and the social theory*. London: Sage Publications.

Part II

Space Policy, Development, and Security in Nigeria

3 Space Politics, Policy, and Strategy in Nigeria

Abstract

Nigeria's space policy and strategy are products of complex dynamics of its national scientific, technological, political, sociocultural and economic developments, with defence and security demands, and international engagements. This chapter examines Nigeria's policy and strategic interests in space, and their social, economic, political, historical and international contexts and determinants. It provides a brief background information on Nigeria's geography, history, state, societies, economy, politics and security as essential introduction to understand the space policy interests, strategies, capacities, benefits and challenges of the country. It examines the political, historical and socio-economic trajectories of Nigeria's space policy and strategy from cultural astronomy in precolonial era to postcolonial spectator at the dawn of space age and then active participant in space research and development. The chapter further reviews extant laws, policies and strategies of Nigeria to determine its national interests and perspectives on space, science and technologies, socio-economic development, and security, as well as their interconnectivity.

3.1 Introduction

This chapter examines Nigeria's space policy interests and programmes in the context of social, economic, political, and historical determinants. For this purpose, this chapter is divided into four sections. The first section of this chapter focuses on the background information about the country, which is necessary to understand the political, socio-economic, historical and security dynamics of its space policy and strategy. The second section examines space R&D in Nigeria in the context of history, politics, economy, and society that shape its development. This is important to understand the third section, which focuses on the policies and strategies of the country on space as well as science and technology, socio-economic

S. Oyewole, *Utilitarianism in Outer Space*, Southern Space Studies,
https://doi.org/10.1007/978-3-031-49646-2_3

development, and security. The fourth section provides the concluding thought on space politics, policy and strategy in Nigeria. Accordingly, this chapter serves as a background to the space policy and strategies of Nigeria, which is important to understand the subsequent discussions on space capacities and capabilities of the country, and their contribution to socioeconomic development and security.

3.2 Background Information on Nigeria

The Federal Republic of Nigeria (FRN) is situated in the west coast of Africa, between 3° and 14° East Longitude and between 4° and 14° North Latitude. It occupies 923,769 square (sq) kilometres (km) (made up of 909,890 km^2 of land area and 13,879 km^2 of water area). Beside its 853 km coastlines in the Gulf of Guinea (along the Atlantic Ocean), its 4,477 km land boundaries are shared with Benin Republic, Cameroon, Chad and Niger (NBS 2017; CIA 2020). The country hosts a network of internal or transnational waters, including several rivers (prominently, Niger and Benue rivers), lakes (prominently, Lake Chad), and lagoon. Beside host of plateaus, Nigeria has large rain forests, creeks, and grasslands, which explain variation in its climate from equator in the south to tropical centre and arid north (CIA 2020; NBS 2017; Ikime 1980).

The entity that is presently known as Nigeria was created through British colonial conquest, occupation and administration between 1860 and 1960. Prior to this period, different pre-colonial political entities that range from small city-states to modest kingdoms as well as large empires and Caliphates occupied and flourished in the region (Ikime 1980). Nigeria became a self-governing or independent political entity on October 1, 1960. It is a heterogeneous country, which has about 250 ethnic groups and over 500 linguistic groups, making Nigeria one of the most ethnically diverse country in the world (NBS 2017; Adenuga et al. 2023). Prominently, the Hausa, Yoruba, and Igbo, which are the three leading ethnic groups in the country, constitute about 61% of the total population (CIA 2020). These among other factors accounted for the adoption of federalism in Nigeria, which is currently divided into 36 states and the Federal Capital Territory (FCT) (Abuja) as well as 768 Local Government Areas (LGAs).

With estimated inhabitants of over 214 million in 2020, Nigeria is the most populous African country and the sixth most populous in the world (CIA 2020). Its population is about 15% of the total population of the 54 African countries (AfDB et al. 2019). Nigeria has a pyramid-form age structure: 41.7% are between 0 and 14 years, 20.27% are between 15 and 24 years, 30.6% are between 25 and 54 years, 4.13% are between 55 and 64 years, and 3.3% are above 65 years (CIA 2020). Amidst these, 62% of adults above age 15 were literate in 2018, which was an improvement on the 51% literacy rate in 2008. Gross school enrolment into primary education was 85% and secondary education was 42% in 2016, while tertiary education was 10% in 2011. In 2006, educational attainment of 1.6% of adults (age 25 and above) was at least master's degree or equivalent (World Bank 2020). The country possessed over 62 million labour force (the 10th largest in the world), of which about 8% are unemployed in 2019 (World Bank 2020; CIA 2020).

Nigeria's Gross Domestic Product (GDP) US$ stood at 397 billion, which was about 17% of Africa's total, in 2018. This made the country the largest economy in Africa and 24th largest in the world (AfDB et al. 2019; World Bank 2020; CIA 2020). The economy is largely based on services, which constituted 56.4% of the GDP in 2017. At the same time, industries contributed 22.6%, agriculture contributed 21.1% and manufacturing contributed 8.8% to the GDP (AfDB et al. 2019). Although oil and gas accounted for about 10% of GDP, it is the major source of government revenue. Government revenue to GDP stood at 7.3% in 2018, making it the lowest in Africa. In 2017, government revenue was 6.6% of the GDP, while its expenditure was 11.8% (AfDB et al. 2019). Accordingly, government has increasingly relied on borrowing to fund its growing expenditure. External debt in 2018 stood at US$ 18 billion, which was about 5% of the Gross National Income (GNI) (Word Bank 2020). Nigeria's GDP per capital is US$2,028 in 2018, making it the 19th in Africa (AfDB et al. 2019).

Nigeria's fourth republic has proven to be its longest years of democratic dispensation. In 1999, when civil rule finally returned to Nigeria, the eight military leaders had ruled for 28 of 39 years of its political independence, leaving the public space militarised (Ojo 2009; Siolun 2013; Oyewole 2020a). In the wake of second republic in 1979, Nigeria abolished the inherited West-minister system for a presidential system of government. It operates a bicameral federal legislature, with the Senate made up of 109 members, drawn on the basis of three senators per state and one from the FCT, and the House of Representative, which has 360 seats (FGN 1999). The tenure of the president and that of the parliamentarians last for four years, and the former is only renewable for two terms. Consequently, elections are periodically held every four years, although their freeness, fairness and peacefulness are debateable (Agbaje and Adejumo 2006; Hamalai et al. 2017; Oyewole and Omotola 2022a). The Nigerian public space is infamous for corruption, state excesses and failure, sectionalism, amoral politics, and militarisation, which are responsible for widespread poverty and deprivation, apathy and discrimination, criminal and political violence, and insecurity (Campbell 2010; Joseph 1991; Nnoli 2008, 2011; Omotola and Alumona 2016; Osaghae 2011; Smith 2007; Oyewole and Omotola 2022b).

The landscape of Nigeria's security has been marked by a series of challenges. Communal, ethnic, religious, and regional conflicts and violence have become defining features of Nigeria's security since 1960s (Adenuga et al. 2023; Nnoli 2008). About seven years into independence, Biafra secessionism led to a 30-months of bloody civil war (1967–1970) that killed over two million (Osaghae et al. 2002; Osaghae 2011). The threats of pro-Biafran secessionists have resurfaced over the last two decades (Onuoha 2013; Oyewole 2019). The rise of oil militancy and criminal violence in the Niger Delta have also threatened Nigeria's economic survival, political cohesion and national security, especially since 2000s (Omotola 2009; Oyewole et al. 2018). The Nigerian-based Islamist Boko Haram insurgents have directly or indirectly killed over 300,000 people and displaced five million in the Lake Chad region since 2009 (Oriola et al. 2021; Oyewole 2015a, 2023c). Lives and properties, road and rail transportation, rural communities and means of livelihood of many in Northern Nigeria have come under the terror of armed banditry (Aina et al. 2023; Ojo et al. 2023, 2024). Nigeria's internal waterways, coasts and the Gulf of Guinea have also been threatened by sea pirates and robbers for decades (Oyewole 2015b, 2016a, 2023d). In addition, Nigeria is a leading hotspot of kidnapping globally (Omotola and Oyewole 2024; Oyewole 2016b). The country is also threatened by sabotage, smuggling, theft of mineral resources, illegal mining and refining, illegal financial flows, cybercrimes and inadequate state security capacities.

It is against this background that Nigeria has developed space policy and strategy to support its desires, capacities, capabilities and efforts toward socioeconomic development and security.

3.3 Space Policy and Strategy in Nigeria: The Historical, Political, and Socio-Economic Trajectories

This section provides some insights into the development of space R&D in the framework of politics and policy in Nigeria. It pays attention to the historical, political, and socio-economic contexts, determinants and factors that characterised the development of space R&D of the country. Accordingly, it serves as a background to the space policy and strategies of Nigeria, which are considered in the subsequent sections of this chapter. For reference, this section is divided into epochs: the genesis and the turning point in Nigeria's space policy trajectories.

3.3.1 The Genesis of Nigeria's Space Policy Trajectories

Space R&D in Nigeria has its distanced origin in astrology and cosmological visions of different pre-colonial cultures and traditions in the country. These are largely characterised by systems of believe, such as myths and rituals, that revolved around the celestial bodies as well as their interactions with terrestrial affairs, including humanity (Chukwuezi 2008; Sègla 2008; Urama 2008). The knowledge produced in this process were largely passed from one generation to another through oral tradition, carving, drawing, painting, music, rituals and myths, and to a limited extent through documented education, especially in the northern part of the country, where influence of Arabic civilisation penetrated and introduced systematic writing. Yet, the knowledge of development as modernisation, industrialisation, monetised economic growth, constructive destruction as well as aggressive human conquest of nature was introduced to Nigeria and most other parts of Sub-Saharan Africa through Western education and European colonisation (Martinussen 1997; Pieterse 2010; Rodney 1972). Amidst these, modern space R&D emerged as a subject of academic interests within pure and applied sciences in Nigeria.

The foundation of future space R&D was laid in Nigeria with the introduction of university and polytechnic, which made advance studies in science and engineering possible, in the country from 1940s (Livsey 2017). Prior these, and after this period, many Nigerians have been trained abroad in relevant fields of space sciences. In 1957/1958, the National Aeronautics and Space Administration (NASA) sponsored the installation of a ground tracking device at the University College of Ibadan, old Western Region of Nigeria, in order to support the launch and operation of the first US satellite *Explorer-1* among other space R&D (Abiodun 2017). In 1960/1961, shortly after the independence, NASA sponsored the installation of a tracking station in Kano, old Northern Region of Nigeria, in order to support Apollo mission, which led to a series of unmanned and later manned spaceflight to the Moon. Nigeria had another major contact with space sciences in 1962, when Zagami—a meteorite from Mars that weigh about 18 kg—landed in the present day Katsina State (Abiodun 2017).

On August 23, 1963, the Nigerian first Prime Minister, Tafawa Balewa, became the first to have the first ever satellite telephone conversation with the then US President, John F. Kennedy. In a conversation that lasted for about two minutes, Kennedy told Balewa that what they were doing was made possible through peaceful use of space, while the latter congratulated the former for the great achievement (US Naval History Blog 2019). Prior to this, there were efforts to create an observatory in Nigeria in 1962, as Profs. Gaustard and Hans Haubolt introduced astronomy and astrophysics to the University of Nigeria, Nsukka (UNN), which was established in 1960. In 1965, a Nigerian, Samuel Okoye (1939–2009) became the first black African to obtain PhD in Astrophysics at Cambridge University, London. He soon returned to Nigeria and joined Faculty of Sciences in University of Ibadan (UI) and later the UNN, where pull of human resources was increasingly developed (Chukwuezi 2008; CBSSA 2020). These among others are indicators of Nigeria's encounter with space R&D at the beginning of space age.

Despite the foregoing opportunities, there was little or no policy interest in space R&D in the first 15 years (1960–1975) of Nigeria's independence. Many factors were responsible for this development. Politically, Nigeria recorded a series of political crises, which undermined policy attention for R&D during this period. These include ethno-regional conflicts, militant party politics, four major (three successful and one failed) military coups, three major secessionist attempts (by the Niger Delta militias, some Northern military officers and the Eastern region or Biafra). Amidst these, the Biafra secessionism led to a three years of civil war, which claimed more than 2 million lives between 1967 and 1970 (Anifowose 2011; Diamond 1988; Nnoli 2011; Osaghae 2011). These made it challenging for Nigeria to develop policy and institutional framework for science and technology, where space policy interests can be nurtured at the time.

Economically, Nigerian policymakers were mostly preoccupied with terrestrial frontiers, such as land and maritime strategies of addressing pressing socio-economic and security questions. Between 1960 and 2008, life expectancy at birth was between 36 and 50 years in Nigeria. The GDP per capital was around US$100 in the 1960s and does not surpass US$ 1000 until 1982. Although the economy relied on agriculture, which also doubled as the largest employer of labour in the country, it was less mechanised. Accordingly, there was less than 1 tractor per 100 square (sq.) km of arable land between 1960 and 1969 (World Bank 2020). Equally, electricity consumption was 23.57-kilowatt hour (KWh) per capital in 1971 (World Bank 2020). These, among others, left the leadership of the nascent country with serious socioeconomic challenges to address. Accordingly, policymakers in Nigeria prioritised agricultural, educational, energy, and industrial policies in the quest for economic development. In this context, attention was only given to scientific R&D that directly support these areas of policy priority, and not space exploration, which as at then had limited economic appeal, especially in most developing countries.

There were also international dynamics to Nigeria's early engagement with space age. Space R&D was considered globally as a luxury, an expensive race for prestige and domination among superpowers and few of their allies who were great powers during the Cold War, especially between 1960 and 1970s. With restricted opportunity for technological transfer, the huge investment that space R&D required made it less attractive to developing countries like Nigeria. Space expenditure of the USA rose from US$1.1 billion in 1960 to US$7 billion in both 1964 and 1965, while that of the USSR was estimated to rise from about US$2 billion in 1960 to US$ 6.7 billion in 1969 (CIA 1966, 1969; Nimmen et al. 1976). Although the two superpowers spent less than five per cent of their GDP on space, such investment was too huge for a country like Nigeria, whose GDP was US$6.6 billion in 1969, thanks to 21.4% growth rate that was recorded from 1968 as against negative growth recorded in the preceding three years. Notably, the proposal for space R&D at the cost of £700 million in Zambia in 1964 failed the test of policy interest for a year-old country with a GDP of $855 million (Oyewole 2017; De Middel 2012). These, among other things, might have informed Nigeria's delayed path in space.

3.3.2 The Turning Point in Nigeria's Space Policy Trajectories

In 1973, Nigeria joined the UN Committee on the Peaceful Uses of Outer Space (COPUOS). However, it was at the 1976's meeting of the Economic Commission for Africa (ECA) and the Organisation of African Unity (OAU) (now African Union- AU) in Addis-Ababa that Nigeria declared interest in space and associated technologies (Abiodun 2017; Akinyede and Boroffice 2013). At the same time, Nigeria displayed serious policy interest in nuclear R&D and considered it weaponization as an option, which was never taken seriously (Adeniji 2000; Lowbeer-Lewis 2010). Many factors were responsible for this sharp departure from the initial indifferent approach of Nigeria to space and related R&D. Prominently, military intervention in politics (starting from 1996), the civil war (1967–1970) and the subsequent rise and fall of Gen. Murtala Muhammed's regime (1975–1976) transformed the Nigeria's national economic, security, and foreign policy dispositions from conservative to radical paths (Ekoko and Vogt 1990; Imobighe 2003;

Tamuno and Atanda 1989). The succeeding Gen. Olusegun Obasanjo's regime (1976–1969) vigorously pursued the policy of economic nationalism, capacity building, and industrialisation with R&D. The aggressive turn of Nigeria's activist foreign policy against neo/colonialism and the apartheid regime in South Africa, which was on the path of developing some capabilities in space, missile, and nuclear R&D as at then, significantly inspired national interests in space amidst considerations for defence, security, and power projection.

In 1976, the Federal Government of Nigeria (FGN) established the National Science and Technology Development Agency (NSTDA), which was transformed into Federal Ministry of Science and Technology (FMST) by Act No.1 of 1980. The ministry was merged with the Ministry of Education (MoE) in 1985 and subsequently regained its independence in 1992 (FMST, 2020). Between 1976 and 1980, Nigeria tried unsuccessfully to launch a centre for remote sensing (Shaba 2012). Thereafter, Nigeria's interests in space received no significant policy attention between 1979 and 1987. The Second Republic under President Shehu Shagari (October 1, 1979 to December 31, 1983) and the subsequent military regime led by Gen. Muhammadu Buhari (1983 to August 27, 1985) displayed little or no interest in space policy.

In 1977, Samuel Okoye secured private funding, and the support of the University of California in the USA, to acquire a 10-m dish operating at 327 MHz for astronomical R&D at the UNN. He also planned a two-station pulsar observation with India and a Very Long Baseline Interferometer (VLBI) observation programme with Germany. However, the effort was not supported by the government, and the dish was never operationalised (Okeke 1999). The lack of policy support for this initiative cannot be fully understood. However, it may be connected to the prevailing political instability in the country, as 1978 and 1979 were the peak of transition to civil rule that was again removed from power in 1983. It may also be explained through the ethno-centric prism of the post-civil war scepticisms among policymakers against the former secessionist eastern region of the country led by the Igbos, especially in R&D with uncertain security implications (Adekanye 1983; Osaghae et al. 2002).

In April 1987, the regime of Gen. Ibrahim Babangida (1985–1993) set up Inter-Ministerial committee to produce the blueprint for the establishment of National Remote Sensing Centre (NRSC), which was approved in 1988, but did not take off until 1996, during the regime of Gen. Sani Abacha (1993–1998). Equally, in 1987, FMST created a National Committee on Space Application (NCSA) with membership drawn from the academic, armed forces and research centres among other relevant government Ministries, Departments and Agencies (MDAs). The National Agency for Science and Engineering Infrastructure (NASENI) was established in 1993 with the mandate that cover space R&D. This Agency subsequently formed a committee of nine-experts, drawn from nationals both at home and diaspora, to develop a draft National Space Science and Technology Policy. These among others paved the way for the creation of National Space Research and Development Agency (NASRDA) on May 5th, 1999, shortly (24 days) before Gen. Abdulsalami Abubakar's regime (1998–1999) finally returned the country back to civil rule, after 25 years of military government (Abiodun 2017; Akinyede and Boroffice 2013; Isoun and Isoun 2013; James et al. 2014; Shaba 2012).

For half century, and particularly between 1976 and 1999, Nigeria failed to substantiate its interests in space with policy reality due to political instability, leadership crisis, inadequate commitment to R&D, policy summersault and economic mismanagement, which plagued every aspect of the society in the wake of twenty-first century (Joseph 1991; Nnoli 2011; Osaghae 2011; Siolun 2013). The reign of the military between 1980 and 1990s also drifted the country toward political and economic failure and collapse as well as international pariah status, to the detriment of internal capacity and external support for space and other scientific R&D. These also forced many talented Nigerians to emigrate to the Global-North for greener pastures. The success of Nigeria's led international mobilisation against the apartheid regime in 1970s and 1980s, and the eventual decline and collapse of the system, equally reduced the drive for catch up and balance of power in space or nuclear capabilities in Sub-Saharan Africa (Adeniji 2000; Imobighe and Ali 2012; Ogwu 2005). Accordingly, Nigeria's status and power were not really threatened, given lack of space capability (satellite in orbit) by its pears in the region until 1998.

There are many factors that drove Nigeria's interests and turned the country into a space player in the early part of twenty-first century. Prominent among these is the return of a passionate space lover and supporter to power, in person of Gen. Obasanjo (rtd) as two-term President between 1999 and 2007. As it will be shown in the subsequent section, he prioritised space among other R&D and considerably mobilised national resources for this purpose, even amidst harsh economic reality of the country (Abiodun 2017; Isoun and Isoun 2013; James et al. 2014). Obasanjo's interests and commitments to Nigeria's space policy and programme were really revolutionary. The succeeding governments of Nigeria, from Umaru Musa Yar' Adua (2007–2010), Goodluck Ebele Jonathan (2010–2015), and Gen. Muhammadu Buhari (rtd) (2015–2023) further sustained Nigeria's space interests and boosted its capabilities. However, many observers are sceptical that the zeal for space has declined in the country (cf. Abiodun 2017). Many experts are worried that government lacked continuity, and progress with space policy has become inconsistently.

The central objective of space policy in Nigeria, as it will be shown in the next section, is to advance socioeconomic development of the country. However, public awareness of the policy and its relevance are limited. Beyond what is publicly acknowledged, however, power politics is equally a salient driver of Nigeria's space interests. Many regional powers that Nigeria would like to consider as peers across the world boosted their capabilities and joined the exclusive space club, by virtue of owning at least a satellite in orbit, toward the end of twentieth century (Abiodun 2008; Harvey et al. 2010; Harding 2013; UCS 2020). This trend became more alarming for Nigeria when African states started to make headways in this regard. The first orbited satellite from the continent was sponsored by Egypt in 1998, and then followed by South Africa in 1999, Morocco in 2001 and Algeria in 2002 (Luncedo, and Gottschalk 2013; Froehlich and Siebrits 2019; Oyewole

2017, 2023a, b). As a regional power, Nigeria cannot avoid being left out (Amusan and Oyewole 2017; Ogunnubi 2014; Ogunnubi and Oyewole 2020; Oyewole 2020b, c). Hence, the country's first satellite was orbited in 2003. These, among other factors, have continued to shape Nigeria's space policy and politics. On these bases, Nigeria's space policy, strategies, institutions, and capabilities will be further examined.

3.4 Space Policy and Strategies of Nigeria

Nigeria's space policy and strategies are products of the politics, history, economy, and sociology of the country, which the last section tried to summarise. Alternatively, they are reflections of the past, present and future efforts and aspirations of the government and people of the country in development and security. Accordingly, the efforts to understand the policy and strategies of the country in space cannot ignore their contexts, most especially the grand strategies and the policy framework. This made it necessary to situate Nigeria's space policy in the context of development and security strategies and extant laws. Hence, this section will be subdivided into two: the first one focus on the constitutional and legislative framework of the space policy, while the second one focus on the details of the strategic policy contents.

3.4.1 The Constitutional and Legislative Framework of Nigeria's Space Policy

Nigeria's interests in space have considerable constitutional backing. The Constitution of Federal Republic of Nigeria (FRN), 1999 makes it a "duty and responsibility of all organs of government, and of all authorities and persons, exercising legislative, executive or judicial powers, to conform to, observe and apply the provisions of Chap. II [Sections 13–24]" on the Fundamental Objectives and Directive Principles of State Policy. Section 14(2b) makes "security and welfare of the people" "the primary purpose of government" in Nigeria (FGN 1999). Therefore, there is no option to development and security for the government of the country in space, given the constitutionally mandated objectives for this and every other issue of policy consideration. Sections 14–24 of the Constitution operationalised the national conception of security and welfare, including national unity in diversity, guaranteed fundamental human rights, political freedom, flourishing civil society groups and private investors, re/distribution of wealth, equality, quality education opportunities, living standard and environment, as well as promotion of national and African interests, while supporting peaceful and just world order in external engagements, among other things (FGN 1999).

Between Sections 14–24 of the Constitution of the FRN, 1999 there are few sections that directly or indirectly justify the space policy and associated commitments. For instance, Section 16(1a) mandates the state to "harness the resources (human and materials) of the nation and promote national prosperity and an efficient, a dynamic and self-reliant economy. In agreement with utilitarian principles, Section 16(1b) states that the state shall "control the national economy in such manner as to secure the maximum welfare, freedom and happiness of every citizen on the basis of social justice and equality of status and opportunity." Section 17(3d) mandates the state to direct policy to ensure that "there are adequate medical and health facilities for all persons." Beside the mandate to ensure "equal and adequate educational opportunities at all levels" in Section 18(1), the succeeding sub-Sect. 3.2 states that "government shall promote science and technology." Section 20 mandate the state to "protect and improve the environment and safeguard the water, air and land, forest and wildlife of Nigeria" (FGN 1999). Space R&D has become an indispensable aspect of science and technology, which are vital to promote quality health, food, environmental and education for all among other constitutional mandates.

In additional to constitutional provision that supported space and related scientific R&D, the legislature has also passed some laws that are important for Nigeria's space policy. Earliest among them is the Nigerian Communication Act of 2003, which established the Nigerian Communication Commission (NCC). Prominent among the responsibilities of the commission is the development and review of national policy on communication; promotion of investment and R&D in communication; as well as management of relevant resources and regulation of communication services (FRN 2003). Although the Act was not explicit with the details, the mandates and the activities of the commission cover regulation and governance of space-based communication technologies and services in Nigeria. The National Information Technology Development Agency (NITDA) Act of 2007 is another important legislation for the space and related technologies in Nigeria. This Act established NITDA and mandated it to oversee and regulate information network and sharing in Nigeria (FRN 2007). Although the Act is not explicit on the details, it makes the agency a critical agency in the management and regulation of information from space-based technologies in the same manner as terrestrial infrastructures.

The NASRDA Act of 2010 is a turning point in space policy legislation in Nigeria. The Act legislated the establishment of NASRDA and allocated 13 functions to the agency. These include responsibilities for building national capabilities in space science and technology (S&T); development of satellite technology, applications, services and launch capacity; conducting space R&D to advance socio-economic development of the nation, coordinate space programmes and their applications; development of strategy for exploitation of outer space and promoting the participation of private sector in the space industry; establishment and supervision of centres that are required for the implementation of national space programme; establishment of information network for the purpose of facilitating communication and data sharing; promotion of active participation of Nigeria in UN-COPUOS and other UN specialised agencies in space S&T; support academic institutions in R&D that are related to space; serving as repository for all satellite data over the territory, as well as collaborator and consultant for all space related data that involve the country; reviewing national policy and development of strategy on space; and finally, every other activities that are expedient and necessary for the operation of the agency and space S&T promotion in Nigeria (FRN 2010).

The Defence Space Administration (DSA) Act, 2016 is another major legislative framework for space policy and strategy in Nigeria. The Act legislated the establishment of the DSA with 15 national policy objectives in mind. These include responsibilities to: (1) facilitate and implement the defence and security aspects of national space policy; (2) develop and operate military space technology; (3) support the Nigerian military operation locally and internationally as well as internal security agencies with satellites; (4) provide affordable and reliable space and cyberspace capabilities for the military as well as other security and law enforcement agencies; (5) support launching of satellites and other space payload; (6) establish space control that is capable of ensuring peaceful use of outer space through the use of navigation, surveillance, protection and space intelligence analysis for counter space operation; (7) establish centres that are capable of providing various satellite application services for force enhancement of the military and the country at large; (8) strengthening the human resource capacity of the military and other security agencies for implementation of defence space programmes; (9) encourage the development and operation of space launch vehicle; (10) coordinate space activities of the services in the armed forces; (11) develop strategies for exploitation of outer space and promote the participation of the military in space industry; (12) collaborate with national and international agencies, centres, private industries and NGOs in space and other relevant fields; (13) promote active participation of Nigerian Armed Forces in UN-COPUOS and other UN specialised agencies in space S&T; (14) support all military formations, commands, units and institutions in space related R&D; (15) conduct sustainable training of relevant personnel, and undertake any other expedient and necessary activities, for the operation of the administration and for the promotion of space S&T in the Nigerian armed forces and other security agencies (FRN 2016).

Besides the foregoing domestic legislative framework, Nigeria has signed and ratified some international conventions and treaties on space and related matters. These include the Outer Space Treaty of 1968, the Rescue Agreement of 1968, the Liability Convention of 1972, and the Registration Convention of 1976 (Eriksen 2019; Momoh 2008; Van Wyk 2008; UN 2002). Although these treaties are legally binding as applicable in International Law, most especially on signatory states, most of them are yet to be domesticated in Nigeria (Eriksen 2019; Momoh 2008). Equally, Nigeria has ratified the Nuclear Test Ban Treaty of 1963, the International Telecommunications Satellite Treaty (ITSO) of 1973, the International Mobile Satellite Organisation Treaty (IMSO) of 1976 and the International Telecommunication Convention of 1992 (Eriksen 2019; Momoh 2008; Van Wyk 2008). However, the country is yet to ratify the Agreement Governing the Activities of States on the Moon and Other Celestial Bodies (the Moon Agreement) of 1979 (Eriksen 2019).

3.4.2 The Strategic Policy Framework of Nigeria's Space Programmes

In 2001, Nigeria adopted a Space Policy and Strategy, which reflected the aspiration of the government and the people as evident in the constitution. Besides the executive summary, the 53 pages' document cover 10 chapters, scope of the national space programme, and an annex on the structure of the NASRDA. As summarised in Table 3.1, the 10 chapters of the document states the goals, objectives, and approaches of Nigeria in space R&D. The policy goals include development of human resources and capacity in space science and technology (SST), as well as applications of SST for poverty reduction, food security, management of natural resources, disaster prediction, warning and mitigation, defence, national security and law enforcement, communication, understanding of Earth atmosphere for national development, education and training, commerce and spin-off benefits and international cooperation (FMST 2021). Several objectives and strategies were identified with each of the goals, although they are considerably repetitive. The last chapter focus on the implementation of space policy. In addition, the policy document provides the scope of the space programme, and in few cases the timeline, and the structure to operationalise it.

The Nigeria's space programmes were guided by the 2021 strategic policy framework for about two decades. In June 2020, the new head of NASRDA called for the revision of space policy and strategy. However, the agency is yet to release any publicly known strategic plan and report of its own. The same thing can be said of the FMST. This is unlike their South African counterparts, which often publish (annual and medium term- two to ten years) strategic plans as well as their assessment report, as it will be shown in this book. Again in early 2020, the NASRDA acknowledged that a proposal was sent to the FGN to revise the budget circle of the agency and is expected to be sent to the parliament if accepted. It was recognised that the agency operates the same financial circle with any other civil service institution in Nigeria, making medium and long term planning quite challenging.

To further underscore the importance of space policy and strategy in Nigeria, there is a need to examine this subject in the context of national development and security policies and strategies. In this way, one can deduce how space policy thrust and strategic goals have been carried through or recognised as a subject of national development and security in Nigeria since 2001. Although the Nigeria vision 20:2020 economic transformation blueprint, which was adopted under President Yar'Adua (2007–2010) with the aim to catapult the country to the list of first 20 economies in the world by 2020, recognised the importance of R&D in science and technology, the 218 pages' document does not identify specific interest in space (FRN 2009). The same thing can be said of the 168 pages' document, Nigeria economic recovery and growth plan (2017–2020), which acknowledged the importance of science and technology generally without any explicit interest in space (Ministry of Budget and National Planning (MBNP) 2017).

Table 3.1 Summary of Nigeria's space policy and strategies

Policy goals	Objectives	Strategies
• Development of human resources and capacity building	Development of: • Poll of Nigerian scientists, engineers and technicians that can design and manufacture hardware and software for various purposes • Skills in various aspects of science and technology • Attractive career opportunities for numerous brilliant Nigerians in STEM • Nigerian technology that can feed local industries and earn-foreign exchange • Highly trained personnel that can train the trainers in STEM	• Funding and resourcing mandatory study of space at all levels of education • Develop space infrastructure for public awareness and curriculum for all levels of education • Enhance the capabilities of institutions of higher learning with space related disciplines • Sponsor satellite supported facilities for distance learning to make educational opportunities available everywhere and avoidable for everyone • Create enabling environment that can attract vibrant Nigerian talents in Diaspora • Establish centre for satellite development, aircraft manufacturing, rocketry and basic space science, among others, to support space R&D

(continued)

Table 3.1 (continued)

Policy goals	Objectives	Strategies
• Poverty alleviation, food security, and management of natural resources	• Agriculture: assessment of crop performance, yield prediction and determine measures to prevent failure • Land resources: assessment, exploitation, and/or management of oil among other resources, pipeline, oil spill and related hazard; terrain and land resources for military undertaking and land use planning; soil for capability and stability for various purposes • Water resources: assessment of the quantity and quality of surface and underground water, maritime resources management, rainfall prediction, drought and other disaster forecast, as well as hydrological systems planning and conservation measures • Ecosystems: Evaluation and monitoring of aquatic ecosystem, land use and vegetation • Plants: Assessment and determination of plant species in forest systems, aquatic plants for consumption, as well as medical and ornament plants • Fauna: Assessment and determination of fish distribution to boost yields of trawlers; areas of overfishing and policy measures against consequences of such; inventory of domestic animals and wildlife; provision of weather information and channel conditions for navigation • Physical environment: Assessment and conservation of air quality and atmospheric indices; provision of short and long terms weather forecast and hazard waring • Education: transmitting teaching material through time and space to impact knowledge	• Enhance the institutions that involved in remote sensing • Establish national ground station for receiving data from satellites • Build and launch Nigerian Earth Observation satellites • Establish equipped laboratories in each state of the federation for satellite data analysis • Adequate funding for the various projected initiated by the space agency

(continued)

Table 3.1 (continued)

Policy goals	Objectives	Strategies
• Disaster prediction, warning and mitigation	• Development of capacity to conduct assessment and monitoring of flooding, land-slides and soil erosion, Whirl wind and thunder storm, heat waves, droughts and desertification, sand and dust storms, locust and other insect infections, bush burning, hazard from oil exploration, waste management, natural or man-made radio activity, impact of climate change on sea level in coaster areas, Earth tremors, inconsistency of crater Lakes, and effects of mining activities, among others, across the country	• Developing enabling environment and integrated mission for collation, analysis and dissemination of satellite data to stakeholders in environmental management • Formulation of national disaster reduction and risk management methodologies, and development of disaster forecasting and preparedness • Development of emergency response system that include disaster, population and resource impacts evaluation and relief administration • Development of recovery and reconstruction plans and actions for the affected populations and infrastructures • Mapping out areas of international cooperation in disaster management
• Defence, national security and law enforcement	The use of space support/ satellite for • Conflict prevention through pre-emptive measures like advanced communication and remote sensing for early warning, border delineation, resource mapping and monitoring of movements of hostiles • Enhancement of military capabilities (command and control for tactical and strategic operations) for the fight against criminal elements internally and externally, regional peacekeeping operations, and protection of strategic interests	• Establishment of space R&D command units in Ministry of Defence (MOD) and its agencies; fund relevant R&D programmes in collaboration with NASRDA, Nigerian Universities and Research Institutions • Production of reliable military communication and navigation; development of software and hardware required by the military in space age, training of personnel in space R&D to meet need of defence and national security • Training for military for disaster operations and participation in search and rescue globally with space support • Employing space capabilities for intelligence tracking and forensic by the police, and in the fight against financial crime advance free fraud

(continued)

Table 3.1 (continued)

Policy goals	Objectives	Strategies
• Understanding the Earth and its environment from space for national development	• Enable and improve understanding or monitoring of the Earth's atmosphere, biosphere, hydrosphere, and magnetosphere, solar-terrestrial relations, climate and weather forecasting for agricultural and socio-economic development, natural resource inventory, pollution (air, maritime and land), ozone layer depletion and climate change, and the impacts for Nigeria, advance warning of natural and man-made disaster; create employment opportunities for young scientists; and produce international cooperation in Earth science	• Encourage the study of basic space science at all level of learning, and the enabling environment to develop space capabilities and socioeconomic benefits in Nigeria • Develop a pool of scientists that can assess upper atmosphere and the effects on human, aircraft and ground services; measure toxic compound over Nigeria and determine its effects for the human, agriculture, forests and natural resources, and proffer effective remedies; understand the impacts of stratospheric ozone depletion and radiation on the population; document available natural resources in the country • Provide ground and space-based monitoring of the Earth and the environment; coordinate national and international monitoring activities to ensure standard and ethical procedure and results
• Space communication applications	• Accelerate economic growth through affordable telecommunication services; increase tele-density of the country in 10 years; link Nigeria with information infrastructure globally • Provide telecommunication services in remote/rural areas; satellite broadcasting among other services like data imaging, multimedia and global internet access; strengthen the development of distance learning education	• Develop and acquire indigenous capacity in satellite and ground receiving technologies; develop and launch affordable and secure satellite communication
• Education and training	• Develop awareness of the socioeconomic benefits derived from space among other fields of science and technology among the Nigerian youths to prepare them for future leadership role • Develop human capacity for R&D of high tech from space R&D for sustainable technological and industrial development; ensure success of high tech and derive benefits from such • Participate in international space R&D, and ensure cooperation with other countries and international organisations	• Revive deteriorating facilities for SST; invest in all areas of space research and training (R&T); sustain policy commitment on space R&T and integrate it into national planning and major projects; ensure mass and continuous education of policy-makers and planners to advance all areas of SST; allocate not less than 5% of education tax to space R&T; provide scholarship for students in different areas of SST; funding of relevant space education centres

(continued)

Table 3.1 (continued)

Policy goals	Objectives	Strategies
• Commercial aspects and spin-off benefits	• Advance from mere consumer to active engagement in research on SST and the spin-off for commercial purposes; attract relevant foreign technologies; create human capacity and employment opportunities for national technological and industrial development; support international cooperation that will promote peace and security of Nigeria and the world	• Provide enabling environment for SST, capacity building, job creation; fund all areas of space R&D; collaborate with established and successful space players with willingness to support the developing countries
• Promotion of international cooperation	Peaceful development and use of space resources for the benefits of Nigerians and all mankind; bi/multi-lateral engagements and cooperation based on mutual respects and complement; development of space capabilities of the country; adopt protective measures on space and its environment and resources through international cooperation; support UN space programmes	support multilateral cooperation on the peaceful uses of outer space in the UN framework; promote with developed and developing nations, sub-regionally in West Africa, and regionally in Africa; support research, business and educational entities to develop international space exchanges and cooperation; build national capacity for bi and multi-lateral collaboration on space programmes, projects and networks, which include ARCSSTE and RASCOM
• Policy implementation	Development of indigenous capabilities in major areas of space R&D, and their utilisation for socioeconomic development	Build capacity in SST, satellite for communications and Earth Observation as well as their applications, indigenous satellite launch vehicles and associated space services in the long run; encourage private sector participation; enhance the education sector in relevant fields; seek international cooperation; establish SST FUND to be managed by NASRDA

Source Table designed by the author with details extracted from FMST (2021)

Despite the foregoing concerns, the 27 pages' policy document on development, known as the Transformation Agenda (2011–2015) that was adopted under President Jonathan identified NASRDA as a critical agency in monitoring and evaluating development programmes of the country (FRN 2011). Moreover, the 80 pages' document, National Security Strategy that was released under President Buhari identified the promotion of R&D in space technology as a subject of defence and national security and restated some of the core goals of space policy in the country (FRN 2019). These, among other things, speaks of the policy commitments and priority for space policy and strategies in the framework of national development and security aspirations and planning of Nigeria.

3.5 Conclusion

This chapter examine the Nigeria's space policy and strategy in the context of relevant historical trajectories and political, social and economic determinants. It provides some insights on the national conceptions of development and security in Nigeria, which all reflected utilitarian orientation of the country. The developmental goals of the government and people of Nigeria, as evident in this chapter, indicate the national interests in economic growth, transformation, catch up, industrialisation, liberalisation and privatisation, resource management, socio-political modernisation, national capacity, human and community flourishing, enlargement of the choices of the people as well as sustainability. The conceptions of security in the country equally reflect elements of national and human dimensions of security, as earlier examined. In addition, the conception of space support for these development and security goals aligns with utilitarian framework of this analysis. Against this background, this chapter is an important basis for the assessment of the contribution of space policy programmes to development and security in Nigeria.

References

Abiodun, A.A. 2008. Future trends in the global arena: challenges and opportunities for Africa. *African Skies* 12: 12–22.

Abiodun, A. A. 2017. *Nigeria's space journey: Understanding its past, reshaping its future.* Abuja: African Space Foundation.

Adekanye, J.B. 1983. Domestic production of arms and the defence industries corporation of Nigeria. *Current Research on Peace and Violence* 6 (4): 258–269.

Adeniji, O. 2000. *Essays on Nigerian foreign policy governance and international security.* Ibadan: Dokun Publishing House.

Adenuga, G., A. O. Olajubu, S. Oyewole, and J. S. Omotola. 2023. Ethno-religious conflicts and the challenges of national security in Nigeria. *South African Review of Sociology.*

AfDB et al. 2019. *African statistics yearbook 2019.* Abidjan and Addis Ababa: African Development Bank Group (AfDB) the African Union Commission (AUC) and the United Nations Economic Commission for Africa (ECA).

Agbaje, A., and S. Adejumobi. 2006. Do votes count? The travails electoral politics in Nigeria. *Africa Development* 31 (3): 25–44.

Aina, F., J.S. Ojo, and S. Oyewole. 2023. Shock and awe: Military response to armed banditry and the prospects of internal security operations in Northwest Nigeria. *African Security Review*. https://doi.org/10.1080/10246029.2023.2246432.

Akinyede, J.O., and R.A. Boroffice. 2013. *Nigeria's quest in space*. Ile-Ife: Obafemi Awolowo University Press.

Amusan, L., and S. Oyewole. 2017. The quest for hegemony and the future of African solution to African development problems: Lessons from headways in African security sector. *Journal of Asian and African Studies* 52 (1): 21–33.

Anifowose, R. 2011. *Violence and politics in Nigeria: The TIV, Yoruba and Niger delta experience*, 3rd ed. Lagos: Sam Iroanusi Publications.

Campbell, J. 2010. *Nigeria: Dancing on the brink*. Ibadan: Bookcraft.

CBSSA. 2020. Our historical background. https://www.nasrdacbss.com/our-historical-backgroud/. Accessed 19 June 2020.

Chukwuezi, B. 2008. The relationship between human destiny and the cosmic forces—A study of the IGBO worldview. In *African cultural astronomy: Current archaeoastronomy and ethnoastronomy research in Africa*, ed. J. C. Holbrook, R. T. Medupe and J. O. Urama, 209–16. Urama. Springer.

CIA. 1966. US and soviet space programme: Comparative size. *CIA Historical Review Programme*, Released, 1998.

CIA. 1969. Expenditure implication of Soviet space programme. *CIA Historical Review Programmee*, Released, 1998.

CIA. 2020. Nigeria. Available at https://www.cia.gov/library/publications/the-world-factbook/geos/ni.html. Accessed 10 March 2020.

De Middel, C. 2012. *The afronauts*, documentary.

Diamond, L. 1988. *Class, ethnicity and democracy in Nigeria: The failure of the first republic*. London: The Macmillan Press Ltd.

Ekoko, A.E., and M.A. Vogt. 1990. *Nigerian defence policy: Issues and problems*. Lagos and London: Malthouse Press Limited.

Eriksen, C. 2019. Nigeria. In *Integrated space for African society: Legal and policy implementation of space in African countries*, ed. Froehlich, A. Springer.

Federal Ministry of Science and Technology (FMST), Nigeria. 2021. *National space policy and programme*. Abuja: Federal Government of Nigeria.

Federal Ministry of Science and Technology (FMST). 2020. Welcome to the federal ministry of science & technology. https://scienceandtech.gov.ng

FGN. 1999. *Constitution of the federal republic of Nigeria 1999*. Abuja: FGN.

FRN. 2003. *Nigerian communication act*. Lagos: FRN.

FRN. 2007. *National information technology development agency act*. Abuja: FRN.

FRN. 2009. *Nigeria vision 20:2020 economic transformation blueprint*. Abuja: FRN.

FRN. 2010. *National space research and development agency act*. Abuja: FRN.

FRN. 2011. *The transformation agenda (2011–2015): Summary of federal government's priority policies, programmes and projects*. Abuja: FRN

FRN. 2016. *Defence space administration act*. Abuja: FRN.

FRN. 2019. *National security strategy*. Abuja: ONSA, FRN.

Froehlich, A., and A. Siebrits. 2019. *Space supporting Africa: A primary needs approach and Africa's emerging space middle powers*. Cham: Springer.

Hamalai, L., S.G. Egwu, and J.S. Omotola. 2017. *Nigeria's 2015 general elections: Continuity and change in electoral democracy*. UK: Palgrave Macmillan.

Harding, R.C. 2013. *Space policy in developing countries: The search for security and development on the final frontier*. New York: Routledge.

Harvey, B., H.H.F. Smid, and T. Pirard. 2010. *Emerging space powers: The new space programs of Asia, the Middle East and South America*. New York: Springer.

Ikime, O., ed. 1980. *Groundwork of Nigerian history.* Ibadan: Heinemann Education Books (Nigeria) Plc.
Imobighe, T. A. 2003. *Nigeria's defence and national security linkages: A framework of analysis.* Ibadan: Heinemann Educational Books (Nigeria) Plc.
Imobighe, T.A., and W.O. Ali, eds. 2012. *Perspectives on Nigeria's national and external relations: Essays in honour of professor A. Bolaji Akinyemi.* Ibadan: University Press PLC.
Isoun, T. T., and M. J. Isoun. 2013. *Why run before learning to walk? Reflections on high technology as a strategic tool for development in Nigeria.* BookBuilders.
James, G., J. Akinyede, and S. Halilu. 2014. The Nigerian space program and its economic development model. *New Space* 2: 23–29.
Joseph, R. 1991. *Democracy and prebendal politics in Nigeria: The rise and fall of the second republic.* Ibadan: Spectrum Books Ltd.
Livsey, T. 2017. *Nigeria's university age: Reframing decolonisation and development.* London: Palgrave Macmillan.
Lowbeer-Lewis, N. 2010. *Nigeria and nuclear energy: Plans and prospects.* Nuclear Energy Futures, No. 11.
Luncedo, N., and K. Gottschalk. 2013. The growth of space science in African countries for Earth observation in the 21st century. *South African Journal of Science* 109 (1–2): 1–5.
Martinussen, J. 1997. *Society, state & market: A guide to competing theories of development.* London: Zed Books Ltd.
Ministry of Budget and National Planning. 2017. *Nigeria economic recovery & growth plan 2017–2020.* Abuja.
Momoh, W. 2008. An overview of Nigerian space activity and space law. *African Skies* 12: 104–108.
NBS. 2017. *Annual abstract of statistics 2016.* Abuja: Nigerian National Bureau of Statistics (NBS).
Nimmen, J. V., L. C. Bruno, and R. I. Rosholt. 1976. *NASA historical data book, 1958–1968 vol. I: NASA resources.* Washington DC: NASA.
Nnoli, O. 2008. *Ethnic politics in Nigeria,* 2nd ed. Enugu: Pan African Centre for Research.
Nnoli, O. 2011. *The struggle for democracy in Nigeria.* Enugu: Snaap Press Nigeria Ltd.
Ogunnubi, O. 2014. Hegemonic order and regional stability in sub-Saharan Africa: A comparative study of Nigeria and South Africa. PhD thesis, University of KwaZulu-Natal, South Africa.
Ogunnubi, O., and S. Oyewole, eds. 2020. *Power politics in Africa: Nigeria and South Africa in comparative perspective.* Newcastle: Cambridge Scholar Publishing.
Ogwu, U. J. (ed). 2005. *New horizons for Nigeria in world affairs.* Lagos: The Nigerian Institute of International Affairs (NIIA).
Ojo, E.O. 2009. Guarding the 'Guardians': A prognosis of panacea for evolving stable civil-military relations in Nigeria. *Armed Forces and Society* 35 (4): 688–708.
Ojo, J.S., F. Aina, and S. Oyewole, eds. 2024. *Armed banditry in Nigeria: Evolution, dynamics and trajectories.* Palgrave Macmillan.
Ojo, S.J., S. Oyewole, and F. Aina. 2023. Forces of terror: Armed banditry and insecurity in North-west Nigeria. *Democracy & Security.* https://doi.org/10.1080/17419166.2023.2164924.
Okeke, P. N. 1999. *Basic space science and technology in Nigeria in the 21st century.* A public lecture organized by the Nigerian Academy of Sciences.
Omotola, J.S. 2009. "Liberation movements' and rising violence in the Niger Delta: The new contentious site of oil and environmental politics. *Studies in Conflict & Terrorism* 33 (1): 36–54.
Omotola, J.S., and I.A. Alumona, eds. 2016. *The state in contemporary Nigeria: Issues, perspectives and challenges—essays in honour of professor Ben O. Nwabueze.* Ibadan: John Archers (Publishers) Ltd.
Omotola, J.S., and S. Oyewole, eds. 2024. *The political economy of kidnapping and insecurity in Nigeria.* Charm: Springer.
Onuoha, G. 2013. Cultural interfaces of self-determination and the rise of the Neo-Biafran movement in Nigeria. *Review of African Political Economy* 40 (137): 428–446.

Oriola, T.B., F.C. Onuoha, and S. Oyewole, eds. 2021. *Boko Haram's campaign of terror in Nigeria: Contexts, dimensions and emerging trajectories*. London: Routledge.

Osaghae, E. 2011. *Cripple giant: Nigeria since independence*. Ibadan: John Archers Ltd.

Osaghae, E.O., E. Onwudiwe, and R. Suberu, eds. 2002. *The Nigerian civil war and its aftermath*. Ibadan: John Archers Ltd.

Oyewole, S. 2015a. Boko Haram: Insurgency and the war against terrorism in the lake Chad region. *Strategic Analysis* 39 (4): 428–432.

Oyewole, S. 2015b. Pirate sanctuary and the political economy of piracy in the Gulf of Guinea. MSc. dissertation, University of Ilorin, Nigeria.

Oyewole, S. 2016a. Suppressing maritime piracy in the Gulf of Guinea: The prospects and challenges of the regional players. *Australian Journal of Maritime and Ocean Affairs* 8 (2): 132–146.

Oyewole, S. 2016b. The fate of hostages: Nigeria's conflict theatres in comparative perspective. *African Security Review* 25 (2): 193–207.

Oyewole, S. 2017. Space research and development in Africa. *Astropolitics* 15 (2): 185–208.

Oyewole, S. 2019. The fracturing of pro-Biafra nationalist movements. *African Conflict and Peacebuilding Review* 9 (1): 1–23.

Oyewole, S. 2020a. Civil-military relations: Conflict and cooperation between military bases and host communities in Nigeria. *African Security* 13 (4): 353–379.

Oyewole, S. 2020b. Military capabilities of regional powers in Africa: Nigeria and South Africa in comparative perspective. In *Power politics in Africa: Nigeria and South Africa in comparative perspective.*, ed. O. Ogunnubi and Samuel Oyewole. Newcastle: Cambridge Scholar Publishing Ltd.

Oyewole, S. 2020c. The quest for space capabilities and military security in Africa. *South African Journal of International Affairs* 27 (2): 147–172.

Oyewole, S. 2023a. The African orbital journey in space age: From spectator to participant. *Vanguardia Dossier* 88 (15): 72–75.

Oyewole, S. 2023b. Military space strategies and African realism: Egypt, South Africa and Nigeria. In *Oxford handbook of space security*, ed. S.M. Pekkanen and P. J. Blount. Oxford: Oxford University Press.

Oyewole, S. 2023c. Piracy and the challenges of Nigeria's maritime domain. *Democracy & Development: Journal of West African Affairs* 7 (1): 55–65.

Oyewole, S. 2023d. Niger's coup weakens regional fight against Boko Haram: four reasons why. The Conversation, August 28. https://theconversation.com/nigers-coup-weakens-regional-fight-against-boko-haram-four-reasons-why-212202

Oyewole, S., and J.S. Omotola. 2022a. Violence in Nigeria's 2019 general elections: Trend and geospatial dimensions. *GeoJournal* 87: 2393–2403.

Oyewole, S., and J. S. Omotola. 2022b. Democracy without demos!: Political mobilization, public opinion and governance in Nigeria. In *Anonymous power: Parties, interest groups and politics in Nigeria* (Essay in Honour of Elochukwu Amucheazi), ed. O. Ibeanu, I. K. Okoye, I. Alumona and E.T. Aniche. Singapore: Palgrave Macmillan.

Oyewole, S., D. Adegboye, and E. Durosinmi. 2018. Militarisation of oil and environmental politics in Nigeria: Armed resistance, state responses and peace prospects in the Niger delta region. *Covenant University Journal of Politics & International Affair* 6 (1): 60–75.

Pieterse, J.N. 2010. *Development theory: Deconstructions/reconstructions*, 2nd ed. Los Angeles: Sage.

Rodney, W. 1972. *How Europe underdeveloped Africa*. Abuja: Panaf Publishing Inc.

Sègla, A. D. 2008. The cosmological vision of the Yoruba-Idaàcha of Benin Republic (West Africa): A light on Yoruba history and culture. In *African cultural astronomy: Current archaeoastronomy and ethnoastronomy research in Africa*, ed. J. C. Holbrook, R. T. Medupe and J. O. Urama, 189–208. Springer.

Shaba, H. A. 2012. Nigerian space programme and African regional perspective.

Siolun, M. 2013. *Soldiers of fortune: Nigerian politics from Buhari to Babangida, 1983–1993*. Abuja: Cassava Republic Press Book.

Smith, D.J. 2007. *A culture of corruption: Everyday deception and popular discontent in Nigeria.* Princeton and Oxford: Princeton University Press.

Tamuno, T. N., and J. A. Atanda. 1989. *Nigeria since independence: The first 25 year. volume IV—government and public policy.* Ibadan: Heinemann Educational Books (Nigeria) Plc.

UCS (Union of Concerned Scientists). 2020. Satellite database. Available at: https://www.ucsusa.org/resources/satellite-database. Accessed on 28 Jan 2020.

UN. 2002. *United nations treaties and principles on outer space.* New York: United Nations Publications.

Urama, E. N. 2008. *Cultural astronomy in African literature.* Master Thesis of Arts in Literary Arts, Department of English and Literary Studies, University of Nigeria, Nsukka.

US Naval History Blog. 2019. Transcript of President Kennedy's conversation with Prime Minister Balewa (JFK Library).

van Wyk, J.A. 2008. Overview of the implementation status of the five united nations treaties on outer space in African countries. *African Skies* 12: 20–28.

World Bank. 2020. World Development Indicators. Available at http://datatopics.worldbank.org/world-development-indicators/. Accessed 10 March 2021.

The Development of Space Capacities and Capabilities in Nigeria

4

Abstract

Nigeria has increasingly developed capacities and capabilities to match its policy and strategic interests in space. In the last three decades, the federal government of Nigeria has established African first dedicated agencies for civil and military space programmes, fully state-owned corporations for commercialisation of space, and numerous centres for space research and development. In addition, few private and non-governmental organisations with exclusive or inclusive interests in space have emerged in Nigeria. These institutions have played major roles in the development and mobilisation of financial, human and material resources behind space access and assets, which are key to the operationalisation of Nigeria's space policy and strategy. To better understand the gap between space policy desires and achievements of Nigeria, this chapter provides an overview of the formation, locations, functions, finances, human and material resources and mobilisation of various governmental, non-governmental and private institutions that are involved in civil, commercial and military space programmes, as well as space advocacies and popularisation, and the resultant space access and assets of the country.

4.1 Introduction

Becoming a spacefaring nation or at least climbing space technology ladder, even with the basic capabilities, have become a sign of national development, and in some cases, an indicator of security for many countries, including the developing ones (Argoun 2012; Harding 2013; Mistry and Gopalaswamy 2012; Wood and Weigel 2011, 2012, 2014). This position is also popular in African space policy studies, where establishment of space agency, publication of space strategy, and sponsoring a satellite has become additional indicator of national development

S. Oyewole, *Utilitarianism in Outer Space*, Southern Space Studies,
https://doi.org/10.1007/978-3-031-49646-2_4

(Abiodun 2008; Froehlich and Siebrits 2019; Luncedo and Gottschalk 2013; Martinez 2012a, b; Munsami and Offiong 2017; Oyewole 2017, 2020). For developing countries like Nigeria, therefore, achieving local and independent capacities in different areas of space R&D are developmental goals that cannot be ignored. As identified in the last chapter, Nigeria's policies and strategies recognised capacity building as a specific development interest that is essential for utilisation of space technologies for broader development and security objectives.

Space policy cannot be reduced to stated objectives and cannot be fully understood without recourse to its operationalisation, that is, the resources devoted to make it a reality. The extent that the objectives have been operationalised equally constitutes a considerable aspect of development. Besides, the contributions of space policy to development and security of Nigeria cannot be assessed outside the context of programmes and capabilities that have been developed by the country in this area. Space programmes and capabilities entail the institutions, activities, assets, human resources, funding and skills among other resources that are made available to operationalise the space policy aspiration of a state. As evident in chapters one to three, these resources are recognised in literatures as elements of national development, and in some cases, human development. Moreover, this section equally offers an opportunity to examine the input, or some of the efforts that are committed to operational realisations of space policy as well as its development and security objectives in Nigeria.

It is against this background that this chapter is further divided into four sections. The first of these sections deals with institutional development for space policy and programmes in Nigeria. The second section focuses on the development of relevant resources that are critical to the mobilisation capacity of established institutions in Nigeria's space policy and programmes. The third section examines Nigeria's space capabilities with emphasis on space access and assets. Finally, the fourth section provides a concluding thought on subject matter of the chapter.

4.2 Institutional Development for Space Policy and Programmes in Nigeria

Institution is an important component of development. Liberalism considers institution as the building block of development. In this case, many students of space policy and politics believe that development of space institution is a key step in the developmental path of space R&D for aspiring and developing nations (Martinez 2009, 2012b, 2016; Shabbir and Sarosh 2018; Wood and Weigel 2011, 2012, 2014; van Wyk 2009). Without relevant institution, a nation cannot be said to have a space programme, as all space activities emanating from such country can be considered as sporadic, unintended, ad hoc and less or un-coordinated for maximum utility. Accordingly, development of space institution cannot be ignored in the assessment of benefits derived by a country in, from and through space.

A nation's space institution broadly covers public (owned by the state or government and non-governmental organisations) and private organisations. While NGOs and private sector can play important roles in a nation's space policy, governmental institutions are critical to create and maintain enabling environment for them to thrive as well as to coordinate their activities among others to advance national interests. In view of these, this section is further divided into two sub-sections: the first one deals with the development of governmental institutions, while the second one focus on NGOs and private sector development, for space in Nigeria.

4.2.1 Development of State Institutions for Space Policy and Programmes in Nigeria

Nigeria's space policy is currently operationalised through three broad-category of space programmes: civil, commercial, and military programmes. In all these cases, the FGN has institutionalised these programmes, as specialise institutions have been developed for each of them. Accordingly, Table 4.1 offers a summary of government space programmes and the institutions developed for these purposes in Nigeria. It is against this background that this section is further sub-divided into three sections. The first one deals with government owned civil space institutions and some of their leading collaborating institutions locally. The second section focuses on the government owned commercial entities, while the third section examine available state institution for military space programmes in Nigeria.

4.2.1.1 Development of State Institutions for Civil Space Programmes in Nigeria

National Space Research and Development Agency (NASRDA) was created in 1999 as the central institution that is mandated to implement and operationalise the Nigeria's space policy and strategy. With this development, Nigeria became the first African country to establish a full fledge space agency (and not centre for remote sensing among other things) and resultantly institutionalised its programmes. Similar agency was only established by Nigeria's regional peers, that is, other leading African nations in space thereafter: Algeria in 2002, South Africa

Table 4.1 State-owned space institutions in Nigeria

Space programmes	Institutions: agencies, corporations and centres
Civil space institutions	NASRDA and affiliated centres: NCRS, CSSTE, CSD, CSTP, CGG, CBSSA, CAR, CBSSA, AAEL, ASTAL, COPINE, ISSE, & ZASTA; autonomous centres: CESRA, SEERL & SWAP Major collaborating institutions locally: UNN, OAU, UNILAG, UNIJOS, FUTA, FUT Minna, RSUST, NMA, UNIBEN, & BUK etc
Commercial space institutions	NIGCOMSAT Ltd; Geo-App Ltd
Military space institutions	DSA and affiliated centres

Source Compiled by the author

in 2005 and Egypt in 2019 (Martinez 2009; Oyewole 2017, 2020, 2023a; Van Wyk 2009). NASRDA has the responsibility to coordinate existing and emerging space institutions, facilities, and programmes in the country. However, this responsibility was technically limited to civil space R&D programmes with the rise of commercial and defence space programmes and associated institutions in the country.

NASRDA is structured into Headquarters (HQs) and development centres (FRN 2010). The Act that established the agency mandated the HQs to be located in Abuja and subdivided into eight departments. These are Office of the Director-General and Chief Executive, Department of Administration and Finance, Department of Engineering and Space Systems, Department of Strategic Space Applications, Department of Policy Planning and Research, Department of International Cooperation, Department of Mission Planning and Satellite Data Management, and the Department of Procurement (FRN 2010; Mohammed 2011). It also empowered the agency to establish more department and division as the need arise.

NASRDA Act 2010 and the Nigeria's space policy acknowledged and mandated the space agency to have six development centres. These include National Centre for Remote Sensing (NCRS), Jos; Centre for Space Science and Technology Education (CSSTE), Ile-Ife; Centre for Satellite Development (CSD), Abuja; Centre for Space Transport and Propulsion (CSTP), Epe; Centre for Geodesy and Geodynamics (CGG), Toro; and Centre for Basic Space Science and Astronomy (CBSSA), Nsukka (FMST 2021; FRN 2010). On NASRDA (2020) website, there are one addition, Centre for Atmospheric Research (CAR), Anyigba. The list of NASRDA centres on CBSSA (2020) website further included three more centres, while the Budget Office (FGN 2018–2020) acknowledged six more centres, which were also given autonomous consideration in budgeting and appropriation. These are Advanced Aerospace Engine Laboratory (AAEL), Oka; Advanced Space Technology Application Laboratory (ASTAL), Uyo; Cooperative Information Network (COPINE), Ile-Ife; Institute of Space Science and Engineering (ISSE), Abuja; as well as the Zonal Advanced Space Technology Application (ZASTA) centres in Langtang, Kashere, and Kano (FGN 2018–2020). NASRDA also has a commercial division known as Geo Apps (Mohammed 2011).

As noted earlier, the NCRS located in Jos, Plateau State is the oldest of all the centres that make up NASRDA. It took off in October 1995. It is primarily charged with the responsibility of managing applications of SST for socioeconomic development, with special focus on the acquisition, archiving, distribution, and utilisation of satellite imageries (FMST 2021; FRN 2010; NCRS 2019). The Centre maintains no fewer than nine departments, including the Directorate, Administration and Finance, Strategic Planning, Geo-Information, Geology and Mineral Exploration, Ground Receiving Station, Hydrology and Climate, Land Resources and Urban Planning, and the Ecology and Environment (NCRS 2019).

Although it appears to be enjoying autonomous consideration in appropriation, COPIN is structure to operate under the coordination of the NCRS as Advanced Space Technology Applications Laboratory-Southwest (ASTAL-SW)

(COPIN 2018). Located in Ile-Ife, Osun State, COPIN has been in existence before NASRDA was established as a parastatal or division of the NITDA. ASTAL, which emerged in 2014, with a base in Uyo, Akwa Ibom State, served similar administrative structures and functions as ASTAL-SW in the Niger Delta region (ASTAL 2020). Similar things can be said of ZASTA centres that emerged in 2017/2018 in Langtang, Plateau State for the North Central geopolitical zone; Kashere, Gombe State for the North-East geopolitical zone; and Kano, Kano State for the North-West geopolitical zone. These initiatives were designed to decentralise space popularisation and applications for socioeconomic development across Nigeria (ASTAL 2020; COPIN 2018). Available details on ASTAL shows that the centres offer consultancy, educational and training services mainly in their areas of jurisdictions. They are structured into six departments. Besides the Directorate, there are Earth Resources Departments (such as the Space Education and Training, the Space Application and Research, the Collaboration and Linkages) and the Decision Support Departments (including the Administration and Finance as well as the Information, Planning and Documentation) (ASTAL 2020).

CSSTE in Ile-Ife, Osun State is also known as African Regional Centre for Space Science and Technology Education-English (ARCSSTE-E). The centre is mandated to develop and train personnel and professional, and develop curriculum, skills and knowledge in Nigeria (FMST 2021; FRN 2010). However, the centre is a product of United Nations General Assembly (UNGA) resolution 45/72 of 1990 and 50/27 of 1995 that the world body should support space R&D in the developing countries (ARCSSTE-E 2020). Accordingly, Nigeria and Morocco were selected to host the UN CSSTE initiative for English and French Speaking African countries respectively. As such, the centre took off in Ile-Ife on November 24, 1998, and was affiliated to the UN-OOSA (ARCSSTE-E 2020; UN-COPUOS 2006). The member states of ARCSSTE-E include mostly English-speaking African countries such as Botswana, Cameroon, Egypt, Ethiopia, Gambia, Ghana, Kenya, Liberia, Lesotho, Malawi, Nigeria, Sierra Leone, South Africa, Sudan, Tanzania, Uganda, Zambia, and Zimbabwe. A report of the UN Committee on The Peaceful Uses of Outer Space (UN-COPUOS 2006) however acknowledged that the centre is placed under the supervision of NASRDA due to the challenges of constituting functioning regional governing board, which made up all member states. Several other sources confirmed the enduring trend in this consideration, as Nigeria continue to manage and fund the institution (ARCSSTE-E 2020; FGN 2020).

CSTP in Epe, Lagos is mandated to develop capacity in rocket science and technology to achieve the Nigeria desire for orbital launch capability (FMST 2021; FRN 2010). Very little is known about this centre and its operation perhaps until recently (Oludimu 2018). The centre has three notable operational units, including Propulsion and Engine System (PES), Structural Integration System (SIS), and Electronic and Simulation System (ESS) (Ibe 2019a). It is equally involved in collaboration and increasingly building network of contact laboratories across the country and beyond, especially in the areas of aerodynamics, chemical thermodynamic, gas dynamic, hydraulic, pneumatic, combustion, electronic, computer, vibration, instrumentation control and guidance among others (Fashade 2008). Closely related to this centre is the AAEL in Oka, Ondo State, which emerged in 2017/2018.

CBSSA was created in 2001 and located at Nsukka, Enugu State. The centre has the mandate to build human and technical capacity in SST with emphasis on astronomy, astrophysics, cosmology, geomagnetism, climatology, rocketry, balloons, and relevant instrumentations for planetary observations (FMST 2021; FRN 2010). The centre is divided into five science departments, including Planetary Science, Radio Astronomy, Instrumentation, ICT and Higher Energy Astrophysics, as well as three non-science departments such as Accounts, Budget and Planning, and Administration (CBSSA 2020). It has developed network of collaboration and maintains several facilities in collaboration with other universities, across the country. The centre is into collaboration with UNN, University of Lagos (UNILAG), University of Jos (UNIJOS), Federal University of Science and Technology, Minna (FUT Minna), Federal University of Science and Technology, Akure (FUTA), River State University of Science and Technology, Port Harcourt (RSUST), and Nigerian Metrological Agency (NMA) among others (UNN 2020).

CAR was created in 2013 and located in Anyigba, Kogi State. It is also known as CAR-NASRDA and NASRDA CAR (NCAR). It has the mandate to understand the Earth's atmosphere, solar physics, space weather and debris among other things, as well as build capacity in these areas and disseminate atmospheric data and products to relevant users for socioeconomic development of the country. The centre is divided into administrative departments, including Account, Registry and Record, and Maintenance; the Technical departments, including Engineering, ICT and Space Research and Initiative; and Organogram (CAR-NASRDA 2020). It has collaborated with many Nigerian universities to develop and maintain a network of laboratories across the country. Prominent among them are Space-Earth Environment Research Laboratory (SEERL) within the University of Benin (UNIBEN) campus and a Space Weather and Atmospheric Physics Laboratory (SWAP) within the Bayero University, Kano (BUK) campus (CAR-NASRDA 2016).

The CGG is located in Toro, Bauchi State. It has the mandate to develop capacity in Earth's survey and mapping, geophysics and geodynamics such as monitoring crustal impacts of drilling and resources exploration, and the rising sea level (FMST 2021; FRN 2010). The CSD is also located in Abuja, FCT. The centre has the mandate to build human capacity in science and engineering for building, launching, tracking, controlling and managing various kind of satellites (FMST 2021; FRN 2010). Closely related to this is the ISSE, Abuja. Unlike CGG and CSD however, it was only created in 2015 and fully established in 2018. The institute is mandated to develop human capacity in SST through training. Although it was established by NASRDA, it is affiliated to African University of Science and Technology (AUST), which was established in 2007 (ISSE 2019). The AUST is the first of the network of centres of excellence to be established under the auspices of the Nelson Mandela African Institute of Science and Technology (NM-AIST), a Pan-African university sponsored by the AU and supported by the World Bank, on the request of many African Heads of State to support the earlier call for this initiatives by the former South African President Mandela (AUST 2020).

Beside the institutions that are directly administered by the NASDRA, there are many other centres across Nigeria. In 2008, NASDRA made a call for application and in the process supported eight universities that are willing to set up their own centre for space R&D. Some of these seeds later became autonomous institutions. Prominent among them are Centre for Space Research and Applications (CESRA), which was sponsored by FUTA in December 2008. The centre is established with the objective of building human capacity as well as conducting R&D in remote sensing and GIS technologies (CESRA 2017; FUTA 2018).

4.2.1.2 Development of State Institutions for Commercial Space Programmes in Nigeria

Development of state-owned commercial entities in Nigeria is an offshoot of the country's civil space R&D programmes. Geo-Apps Plus Limited and the Nigeria Communication Satellite (NIGCOMSAT) Limited emerged from NASRDA as the major public corporations in Nigeria's space sector. These made Nigeria one of the very few African countries to have developed state-owned space institutions and programmes for commercial purposes.

Formally known as the NigeriaSat Imageries and Consultancy Services Limited, Geo-Apps Plus Ltd was established in 2003 and made a commercial arm of the NASRDA in 2007. The corporation offers consultancy services and sell data generated from Nigerian Earth Observatory (EO) satellites to relevant buyers (Space in Africa 2019b). As evident in the industrial report of Space in Africa (2019b), this corporation is the oldest and perhaps the only state-owned entity in Africa that commercialise EO data from satellites sponsored and operated in the continent. Other entities in this area are mostly privately owned.

NIGCOMSAT Limited was incorporated as communication satellite service provider in Nigeria in 2004. It was initially under NASRDA in FMST before the struggle for autonomy by the leadership of the corporation led the FGN to move it under Federal Ministry of Communication Technology (FMCT), where it is operating with the regulation of the NCC (Daily Champion 2011). The corporation also further pushed for its own law, NigComSat bill, which has failed to secure enough support in the National Assembly on different occasions. The bill has been resisted by different stakeholders, including ministries (like FMCT and FMST), agencies (NASRDA and NCC), and many in the private sector in the same industry that believed it will put them in comparative disadvantage (Osuagwu 2012). Instead, many have advocated for full privatisation of the corporation, and Bureau of Public Enterprises (BPE) have made efforts toward this direction at different points in time but was retrained by advocates of national security (Adebajo 2020; Francis 2019; Elebeke 2019; Space in Africa 2020a). NIGCOMSAT Ltd. is the first fully state-owned corporation with communication satellite in Africa—its closest predecessor is NILESAT, which was established in 1996 as a public–private partnership that involved the Egyptian government as one of the shareholders (Space in Africa 2019b).

4.2.1.3 Development of State Institutions for Defence Space Programmes in Nigeria

Nigeria established the Defence Space Agency on October 9, 2014, although it was later legislated as the Defence Space Administration (DSA) by the National Assembly in 2016 and was signed into law by the President Muhanmadu Buhari on February 3, 2017 (DSA 2020). The administration is mandated to operationalise the defence and security aspects of Nigeria's space policy and programmes (FRN 2016). This development has increasingly streamlined the primarily role of the NASRDA to civil space programmes. Although there are indications that South Africa, Egypt, Libya, Algeria and perhaps Morocco are involved in military space activities prior to this development, this initiative made Nigeria the first African country to develop a publicly known specialise institution for its defence space programme (Oyewole 2020, 2022, 2023a, b).

Besides two separate offices for the Chief Executive Officer (CEO) or the Chief of the Defence Space Administration (CDSA) and the Deputy Chief of the Defence Space Administration (DCDSA), the institution has nine directorates, including Directorate of Earth Observation (DEO), Directorate of Satellite Communication (DSC), Directorate of Navigation, Positioning and Timing (DNPT), Directorate of Launch Services and Space Operations (DLSO), Directorate of Policy, Planning and Research (DPPR), Directorate of Cyber Security (DCS), Directorate of Support Services, Directorate of Finance and Account (DFA), and Directorate of Procurement (DP). DSA was also mandated to operate five centres: Space Innovation and Development Centre (SIDC), Defence Earth Observation Centre (DEOC), Defence Satellite Communication Centre (DSCC), Defence Cyber Operation Centre (DCOC), and the Defence Space School (DSA 2020; FRN 2016).

4.2.2 Development of Non-state Actors (NGOs and Private Sector) in Nigeria's Space Policy

NGOs and private sector are important in the consideration of institutional development for space programmes. They are critical to the political and economic liberalisation of space and associated benefits, as essential platforms for wider education or popularisation, mobilisation, and participation in space policy processes. Their expansive services also make space, its technologies, and associated benefits more accessible and, in competitive environment, more affordable to greater majority of the world population. Accordingly, the development of such important institutions, platforms or sector cannot be ignored in the assessment of Nigeria's space policy, its operationalisation and their contributions to the development and security of the country. Table 4.2 offers a summary of known space-related NGOs and private companies in Nigeria as of 2020. This section is subdivided into two: development of NGOs and private sectors in Nigeria's space policy and industry.

Table 4.2 Space-related NGOs and private companies in Nigeria

Major non-state space actors	Major organisations
NGOs	SGAC, EarthSpace, AWB Nigeria, NAS, GSN
Private Corporations	Space in Africa, Atlantic Factorial, BeepTool

Source Compiled by the author

4.2.2.1 Development of NGOs in Nigeria's Space Policy Processes

Nigeria has few known NGOs that make space their core and perhaps only areas of interest. Prominent among them is the Nigerian Astronomical Society (NAS). It is an umbrella body for professionals that are involved in teaching and research in astronomy, space science and other related disciplines (Space in Africa 2018a). The organisation promotes space science education and research through teaching and engagements with academic community, university students inclusive, as well as public enlightenment and mobilisation. It maintains a strong connection with CBSS among other government space institutions and facilities as well as relevant NGOs and private firms. It holds annual conference, which started in 2011.

The national branch of the Space Generation Advisory Council (SGAC) is another major NGO with space focus in Nigeria. The Council is an INGO with the aim of representing university students and young space professionals between 18 and 35 years of age and their interests in engagements with the UN, space agencies, industry, and academic community (SGAC 2020a, b). Established in 1999, the organisation has HQ in Vienna and branches in more than 150 different countries, including Nigeria. It enjoyed a permanent observer status at the UN COPUOS as well as its Technical Subcommittee (TSC) and Legal Subcommittee (LSC), and consultative status at the UN Economic and Social Council (UNESC) (SGAC 2020a, b). Nigerian representations in the organisation dated back to its formative year. The representation of the country is headed by a programme manager, which work in collaboration or to assist the regional coordinator, and is answerable to the executive committee of the organisation. The national and regional branches of the Council have strong relations with the NASRDA, especially at its formative years.

EarthSpace is a Lagos-based NGO with primary focus on space in Nigeria. It was founded by Dr Fayetola Ayodele and incorporated in Nigeria in 2003 as a limited liability company, but maintain a sort of NGO (foundation) arm. A review of the activities of the organisation simply suggests that it operates more like an NGO than a private company. Prior to its development, Mr Ayodele initiated the Space Science Popularisation Programme in Nigeria (Space SPiN), between 2001 and 2007, in collaboration with Babatunde Rabiu, now Professor and Director of one of NASRDA's centre. The primary objectives of both initiatives are to popularise space through training and educational outreach. The organisation has collaborated with the National Geographic (NG), the American Geophysical Union (AGU) as well as some American-based universities and Lagos-based telecommunication companies.

The Astronomers Without Borders (AWB), Nigeria was established in October 2013 and incorporated as an NGO under Companies and Allies Matters Act (CAMA) of the Corporate Affairs Commission (CAC). The organisation was founded by Olayinka Fagbemiro with the aim of promoting STEM and making astronomical observations accessible to all Nigerians (AWB 2020). The organisation is affiliated and in partnership with Office of Astronomy for Outreach (OAO), Office of Astronomy for Education (OAE), Office of Astronomy for Development (OAD), International Astronomical Union (IAU), West African Regional Office of Astronomy for Development (WAROAD), AWB International, NASRDA, DSA, Mount Wilson Observatory (MWO), Los Angeles Astronomical Society (LAAS), African Astronomical Society (AAS), Sterren Schitten Voor Ledereen (Stars Shine for Everyone) (AWB 2020). The office of the organisation in Nigeria is located in Abuja, FCT.

There are many other NGOs with implicit focus on space. Prominent among them are the Nigerian Meteorological Society (NMETS), the Geoinformation Society of Nigeria (GSN), the Nigerian Society of Physical Sciences (NSPS), and the Nigerian Society of Engineers (NSE). The NMETS was established in 1986 as an umbellar body for Nigerian professionals who are involved in teaching, research and public campaigns on weather and climate. The organisation is strongly connected to the Nigerian Meteorological Agency (NIMET), the World Meteorological Organization (WMO) and the NASRDA (NMETS 2011; TheNews 2019). The GSN is the foremost Nigerian professional association on geospatial analysis, which is established with the aim of educating the public on, and promoting the, subject matter (GSN 2020). The NSPS was incorporated in Nigeria on January 9, 2019. It was established with the aim of promoting research and study of physical sciences in Nigeria and develop effective programmes to support this objective (NSPS 2020). The NSE was established on February 16, 1958, and incorporated in Nigeria in 1967. The organisation works closely with the government, academic community, and industries with the aim of setting engineering standard, develop human capacity as well as facilitate and promote technological adaption, acquisition, and development in Nigeria (NSE 2020).

4.2.2.2 Development of Private Sector in Nigeria's Space Industry

Nigeria has very few known private companies in space industry. Prominent among them are Space in Africa, which was founded by Temidayo Oniosun in 2018. Mr. Oniosun is a meteorology graduate from FUTA, the former African regional coordinator for SGAC, and a young entrepreneur in his 20s. With HQs in Lagos, the organisation has developed physical presence in five African countries. Space in Africa is a media company that provide news as well as market and business assessment of space industry in the continent, covering 54 countries (Space in Africa 2019a). It is into partnership with the AU and various national space agencies, NGOs, and private companies across the continent and beyond.

Other major space related companies in Nigeria are Atlantic Factorial Limited (AFL) and BeepTool. Atlantic Factorial was established as a limited liability company in 2010 with a base in Lagos. It uses satellite data among others to provide tracking services to public and private establishments in Nigeria. The company is in partnership with the NASRDA (AFL 2020; Space in Africa 2019b). BeepTool Communications and Integrated Ltd. was founded by John Enoh in 2014. The company deals with affordable smart satellite phone. It partners with the Avanti Communications, Kepler, NigComSat, OneWeb, Telesat, and Sky and Space Global (Space in Africa 2019b; BeepTool 2020). The Lafiya Telehealth App Ltd. is one of the popular subsidiaries of BeepTool group of companies (Space in Africa 2019b).

4.3 Development of Mobilisation Capacities for Space Policy and Programmes in Nigeria

Institutions cannot operationalise space policy and strategy beyond the resources that are made available to them. Human and material resources are required to develop the capacity of an institution in the efforts to build space capabilities of a country for development and security. However, the development of these resources and their management are essential part of any organisation as well as its success and failure in delivering its objectives. This section therefore examines the human and material resources that are available to achieve the development and security aspirations of the people and government of Nigeria in space.

4.3.1 Human Resource Development in Nigeria's Space Sector

Human resource is the most important aspect of every organisation. It involves the quality and quantity of personnel, workforce and labour that make up an organisation. The quality of personnel includes the skills, commitments, procedure, and motivations of personnel that make high performance workforce (Noe et al. 2011; Senyucel 2009; Sims 2002). This implies that the quantity of personnel is one aspect of human resources, the quality of their training and skills as well as material wellbeing and resources made available for job satisfaction and motivations are another thing. Some of these elements will be considered here, while later discussion on material resources will provide more insight on the subject. Besides, this subject of concern is equally relevant to human capital development, which constitute one of the contributions of space policy programmes to the general human development in Nigeria.

This study is unable to determine the exact number of human resources that are available to operationalise the Nigeria's space policy. A report by Space in Africa (2019a) shows that 8,500 people were employed in space industry, including public and private facilities, across the continent. Amidst these, over 3,000 personnel are employed by NASRDA (Space in Africa 2019a). This amounted to over 35% of continent's space workforce. In 2018, NASRDA was reported to have sponsored 600 Nigerians in acquiring postgraduate degrees, including 200 PhD and 400 Master degree holders, in different fields of space sciences and engineering, and planned to train up to 1,000 persons by 2025 (Ibe 2019b). Moreover, a senior official of the agency disclosed that the institution has up to 200 PhD holders trained across the world, and over 400 currently on training as of 2020. Another source within the agency claimed that the agency has up to 300 PhD holders, which was believed to be higher than that of many universities in the country. Nevertheless, another internal source claimed that the agency has over 180 PhD holders, which make it the highest concentration of such degree in any government agency and department in the country.

However, some sources outside the agency recognised that government space institutions are considerably overstaffed, in the sense that they have more personnel than the facilities that are available for them to work. Amidst these, there are concerns about how politicians have influenced the recruitment processes to undermine merits and flooded some of the institutions with their candidates, which are in possession of varying degrees of competence. However, a source revealed that this development is recent, given how recruitment into the agency was very rigorous with several stages of test, which only produced the very best, during the formative years of the agency. The politicisation started with some elements of ethnic, religious and regional mobilisation, as the subject of leadership succession of the institutions arose. In addition to these are the growing issue of party loyalty, especially for the leadership. These and the concerns associated with resource management will receive more attention in the next section on material resources that have been committed to space in the country.

Beside NASRDA and affiliated centres, there are few other actors in Nigeria's space industry with their own personnel. NigComSat was reported to have between 201 and 500 personnel (Space in Africa 2019b). Amidst these, 50 personnel were trained abroad in different areas of satellite communications, and they have in turn trained over 100 more. These are in addition to supporting staffs. At the same time, GeoApp Ltd was reported to have 58 staffs (Space in Africa 2019b). Equally, the personnel strength of DSA can be estimated to be less or around 100 staffs. These are made up of some civilians with required expertise, who were recruited directly or transferred from NASRDA, as well as specially trained personnel drawn from the Nigerian Army (NA), Nigerian Air Force (NAF), Nigerian Navy (NN) and the Nigerian Police Force (NPF). However, military and paramilitary personnel are deployed or transferred in and out of the DSA regularly. Hence, the number of those that have trained and worked with the administration are hard to determine.

Private companies employ lesser workforce in Nigeria's space sector. As at 2019, AFL and BeepTool were reported to have employed 15 and 13 staffs respectively (Space in Africa 2019b). Space in Africa operated with 10 to 20 staffs within and outside the country. EarthSpace utilises less than 10 ad hoc staffs and volunteers, mostly students that are trained and paid stipend, whenever it has a major programme.

The membership strength of space-related NGOs can also offer some insights into human capital development in this sector. SGAC has more than 15,000 members in more than 150 countries (SGAC 2020a, b). Amidst these, a source that is familiar with organisation claimed that Nigeria has close to a thousand members, and was at a point in time the 5th largest and most represented country in the organisation. These are in addition to the membership strength of AWB Nigeria, NAS and GSN among others.

Despite the foregoing, the staff strength of these institutions is not enough to determine the human resources that are operationalising the Nigeria's space policy and strategy or that are available for this purpose. There are many lecturers and researchers with PhD as well as postgraduate and undergraduate students across hundreds of higher institutions (universities, Polytechnics, Colleges of Education and Technical Colleges), and graduate, in different areas of sciences and engineering, in Nigeria. For instance, more than 64,000 students were admitted into various engineering courses across Nigerian universities in 2015 alone (NBS 2017). These are in addition to tens of thousands of Nigerian scientists, engineers and technicians in diaspora. In essence, Nigeria has huge human resources that can be mobilised to actualise its aspiration through space among other subjects of concern in science and technology.

4.3.2 Material Resources in Nigeria's Space Sector

Although human resources are the most important aspect of any organisation, the output or performance of the workforce largely depend on the material resources, including funding, facility, equipment, and work environment that are made available for their operation. The contribution of space policy to development and security in Nigeria cannot be fully realised or assessed without paying adequate attention to operationalise all the relevant elements of the inpute, including the material resources that are available to the institutions and the associated workforces that are commissioned to implement the policy and its programmes. In this consideration, this section will pay attention to the available funding and facilities.

4.3.3 Space Expenditure in Nigeria

Funding is a critical aspect of any organisation and determinant of its institutional capacity. Nigeria has committed hundreds of billions of naira to its space programmes since its inception. Inadequate access to the full annual reports of Euroconsultant and Space in Africa make it challenging to determine the space spending of the country. However, Nigeria often appear in Euroconsultant's global map of countries spending US$10 million and above on space. Some of the available details from these reports show that Nigeria's space spending was US$70 million in 2013, US$66 million in 2014, US$61 million in 2016 and US$48 million in 2018 (Bochinger 2016; Euroconsultant 2017; Lal et al. 2015; Seminari 2019). In this way, Nigeria's space spending was the second largest in Africa (after Morocco) in 2013, first in the continent in 2014 as well as fourth (after Algeria, Morocco, and Angola) in 2016 and (after Egypt, Morocco and Algeria) in 2018 (Oyewole 2020). While the methodological details behind Euroconsultant data is not fully available for proper scrutiny, data from the Nigerian budget office and Appropriation Act, as well as their conversion to the US$ from naira (N) using the average exchange rate for each year as documented by the World Bank's World Development Indicators, questioned the figures given on space budget of Nigeria. This will receive further attention shortly.

In 2018, a Business & Market Analysis by Space in Africa (2018b) shows that Nigeria voted N13.7 billion (US$37.8 million) to civil space programmes, which was 18.1% of the total budget of the FMST. The same report shows that Nigeria budgeted N2.9 billion (US$8 million) on military space programmes, which was 0.5% of the total military budget. However, the list that captured the breakdown of Nigeria's civil space institutions and their budgets only cover 11 of them and excluded CSTD and COPINE. The report also ignored N5.169 billion (US$16.89 million) budgeted for NIGCOMSAT Ltd, which is government owned commercial space programme, in 2018. These among other things questioned the authoritativeness of the given quantification of Nigeria's space spending by the report. The limited access to the data of Euroconsultant and Space in Africa as well as the inconsistency of the available ones make it challenging to understand the history or draw trend analysis of financial resources that have been committed by the government of Nigeria and perhaps the few existing NGOs/private corporations to different aspects of space programming.

Nigerian Budget Office archived FGN budget on its website since 2009. Amidst these, an independent effort was made to track the budget allocation to different space institutions in Nigeria in the last decade. Figure 4.1 provides a summary of the findings. Nigeria budgeted N137.987 billion (US$640.6 million) to its space programme between 2011 and 2019. It is amidst these that NASRDA HQ and NIGCOMSAT received the lion share of 28% and 24% respectively. Some of the long-established institutions like NCRS, CSTD and CSTP also received significant share of the space funding with 8.69, 7.96 and 6.55%. It is also evident that the newly established DSA is receiving significant budget attention, as it accounted for 15.4% of the Nigeria's space spending in 2019. However, it is important to note that Nigeria's space spending is insignificant in the total government spending, as it accounted for 0.034% in 2019. Moreover, it is notable from Fig. 4.1 that space spending always fall during the election years from what it was at the preceding years. This is the case during the 2011, 2015 and 2019 election years.

Despite the foregoing, there are many concerns against known details of Nigeria's space expenditure, which is often gaged based on budget. Prominent among these concerns are the fact that not all budget allocations are usually release on time or utilise judiciously. It was learnt that sometimes the agency doesn't receive funds for necessary projects in early months of the year, due to regular delay with appropriation processes and release of fund for capital projects in the country. In this way, many staffs are often redundant and project schedules are often disrupted. More alarming, unutilised fund by ministries, agencies and parastatals have to be returned back to treasury at the end of the year. Hence, a project that ought to run for 10 months, and the funding is release later than the first two months of

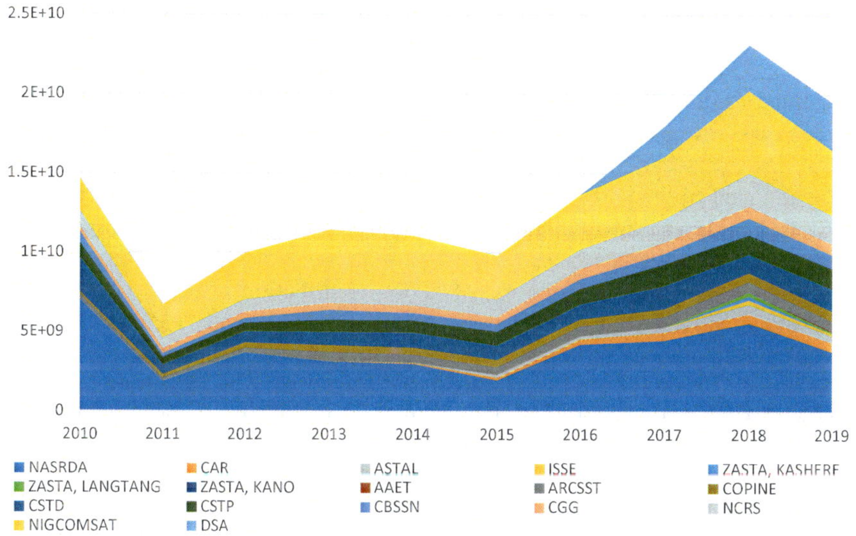

Fig. 4.1 Nigeria's space budgets, 2011–2019 (in Naira). *Source* Data extracted from Budget Office (2010–2019)

the year would see the remaining money return back to FGN's cover and reallocated for the next year. These among other things delayed many projects and give room to diversion of funds. Beside some of the concerns raised by few contacts, there are some publicly available records of allegations and counter-allegations of corruption, diversion and mismanagement of fund, and non-performing allocation and investment that are associated with the Nigeria's space programmes (Adebajo, 2020; Anderson 2007; Blueprint 2020; Pen Pushers 2020; Punch 2016; The Nation 2018; The Whistler 2019; Sahara Reporters 2020; Space in Africa 2020b).

Most of the Nigerian space NGOs are funded with grants, donations, aids and sponsorships from local and international organisations. Accordingly, AWB, Nigeria has been supported by OAD, AWB International, IAU, UNAWE, VIXEN company, SSVI Telescope, and the Nigerian Army, among others (Fagbemiro 2020). The EarthSpace has also received similar international grants and donations, most especially from the USA. Professional associations like NSA, GSN and NSPS rely on annual dues from members and support from relevant government agencies and institutions. However, private companies rely on market forces and foreign investment as sources of their funding (this will be considered in details later).

4.3.4 Ground Facilities and Infrastructural Development for Space R&D in Nigeria

Ground facilities and related infrastructural capacities for space R&D can be considered as an aspect of infrastructural development for national growth, development, and security. Facility is a critical aspect of material resources and major determinant of the performance of the workforce in relation to the targeted policy goals. Beside allocated funding, the facilities that are available for the operationalisation of the Nigeria's space policy cannot be ignored in the consideration of key capacities that determine the extent of the contribution of the subject matter to development and security in the country. Facilities generally include land, building, and equipment. In this consideration, this sub-section briefly examines the facilities of some of the institutions that operationalise the space programmes in Nigeria.

Figure 4.2 broadly shows a satellite imagery of Nigeria's space facilities in Abuja, the FCT. These include headquarters (HQ) of the NASRDA, together with CSTD, ISSE and DSA, which are all located in a property that is introduced as Obasanjo Space Centre (OSC), with close proximity to NIGCOMSAT. Figure 4.3 provides a closer view of NASRDA HQ, DSA, CSTD and ISSE. This property was disclosed to host a ground receiving station for the Earth Observatory satellite inventory of the country. Also, it was supposed to host a facility for Satellite Assembling, Integration and Testing (AIT). Although this project has received budgetary attention running into hundreds of millions of naira over the years, it has not been completed or equipped as of 2020. While there is public insinuation of diversion of fund on the project, it has also been revealed by insiders that government has not release enough fund to complete the project (also see The Whistler 2019). Although the DSA HQ is located in Abuja, there are evidence that they are constructing other facilities elsewhere. The property is also in the proximity of NIGCOMSAT HQ (see Fig. 4.4) and a host of facilities, including a ground receiving station for communication satellite.

Figure 4.5 offers a satellite imagery of the NCRS facility in Jos. The facility hosts remote sensing satellite ground receiving station for utilisation and distributions of satellite imageries to relevant users as well as archiving for socioeconomic development. The centre has received significant budget attention as evident in the last section. Although this suggest that the centre has developed appreciable capacity, the quality of the facility cannot be fully determined. Figure 4.6 is a satellite imagery of ARCSST, which is located within the campus of OAU in Ile-Ife, Osun State. Figure 4.7 also provides satellite imagery of CBSSA, Nsukka, Enugu State. The centre host facilities that include astronomical observatory, which is equipped with 0.25-m and 0.15-m telescopes, as well as radio telescope (UNN 2020). There were indications that the CSTP have equally completed the construction of its facility with fix and mobile launch pads for rocket.

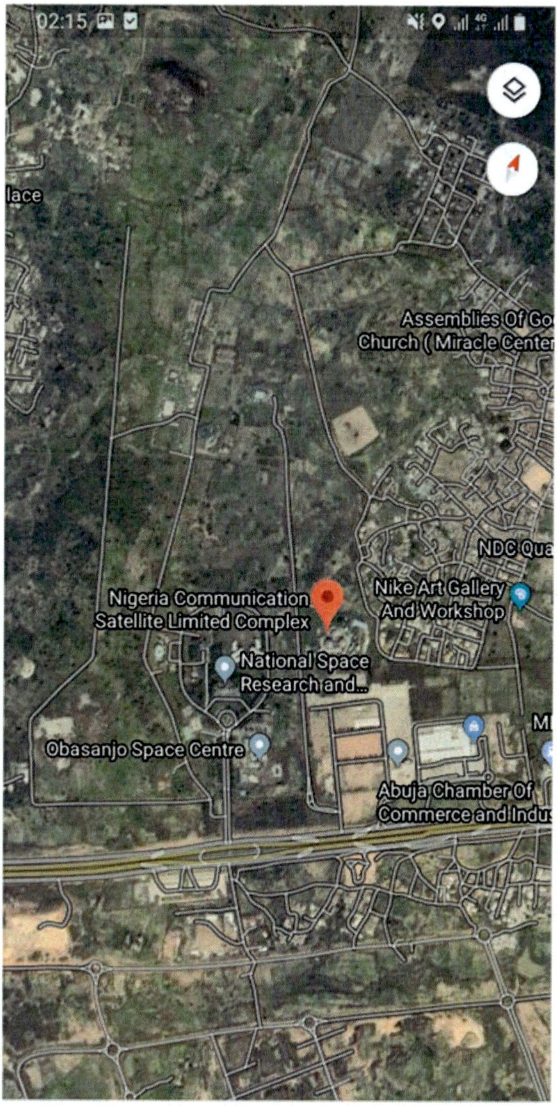

Fig. 4.2 Satellite imagery of Obasanjo Space Centre and NIGCOMSAT, Abuja, Nigeria. *Source* Tracked and captured by the author using GPS, Google Map (2020)

The agency is also affiliated with many other agencies and centres as well as universities, which also maintain facilities that are directly or indirectly relevant to space R&D. Prominent among these is the satellite ground station owned and operated by the CESRA in Akure. These are in addition to international cooperation maintained with different countries across the continent and beyond. South

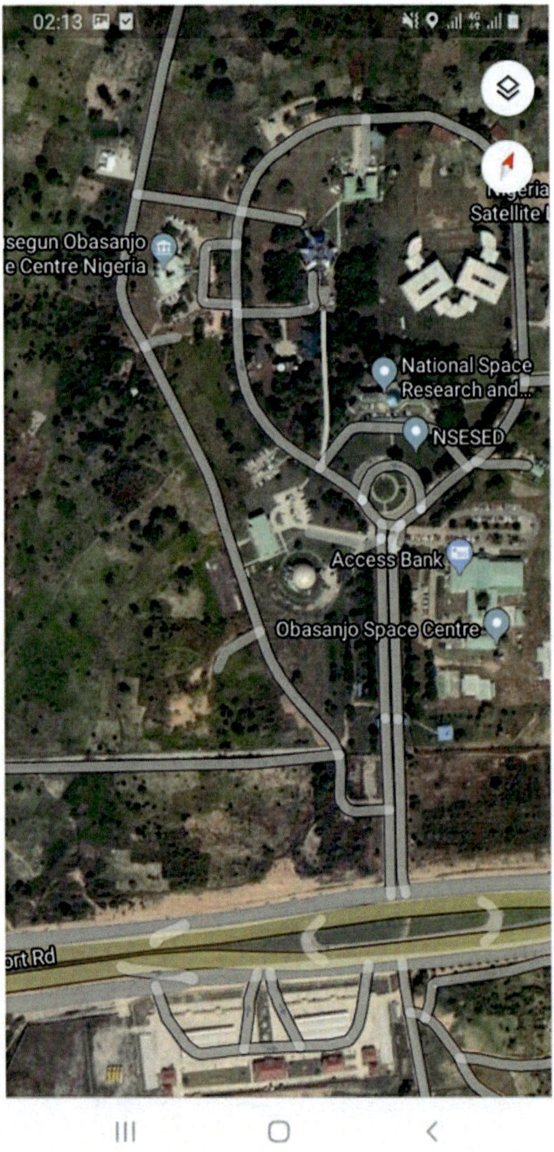

Fig. 4.3 Satellite imagery of Obasanjo Space Centre, NASRDA & DSA, Abuja. *Source* Tracked and captured by the author using GPS, Google Map (2020)

Africa, the UK, China, Japan and Russia are prominent among the partners that are engaged by Nigeria, of which some of them have opened their facilities to the latter for human capital development, services and other collaborative space activities.

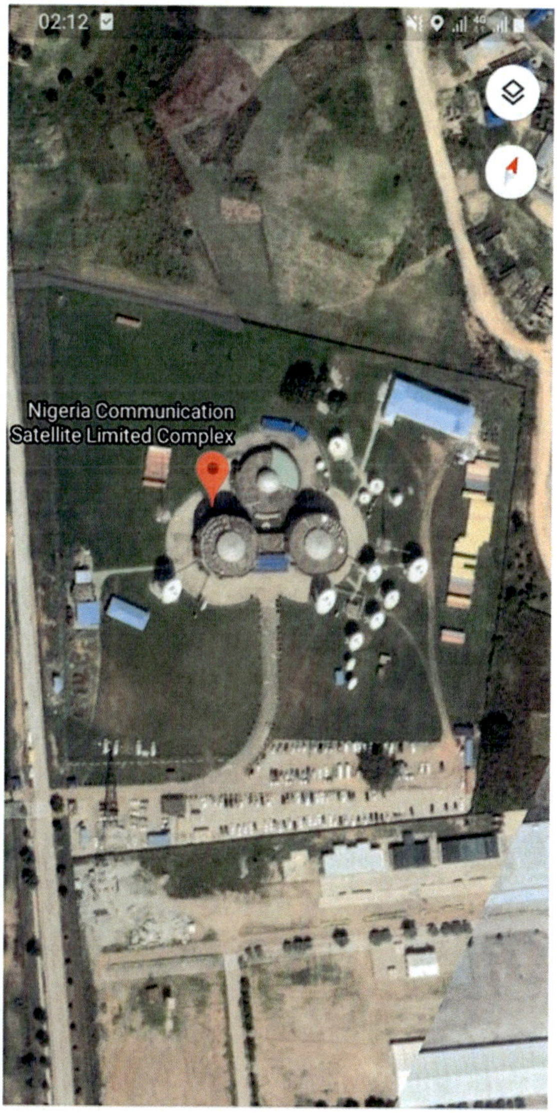

Fig. 4.4 Satellite Imagery of NIGCOMSAT facility, Abuja. *Source* Tracked and captured by the author using GPS, Google Map (2020)

4.4 Development of Space Access and Assets in Nigeria: Satellites and Launch Capabilities

With or without a space agency, sponsoring, owning and operating at least a space asset, that is, a satellite in orbit, and developing capacity for space launch, are considered as major steps that qualify a nation as a spacefaring country (Argoun 2012; Harding 2013; Othman 2003; UCS 2020; Wood and Weigel 2011, 2012, 2014). Accordingly, the progress made by Nigeria in satellite and space launch

Fig. 4.5 Satellite imagery of the NCRS facility, Jos, Nigeria. *Source* Tracked and captured by the author using GPS, Google Map (2020)

capabilities are important in this consideration. To address these concerns, this section is sub-divided into two: the first one deals with the development of space-based capabilities in Nigeria, while the second one focus on the nation's space launch capabilities.

Fig. 4.6 Satellite imagery of the ARCSSTE-E facility, Ile-Ife, Nigeria. *Source* Tracked and captured by the author using GPS, Google Map (2020)

4.4.1 Development of Space-Based Capabilities in Nigeria: Satellites

Nigeria has developed significant space-based capabilities by African standard. The country has sponsored or co-sponsored the orbiting of seven satellites as evident in Table 4.3. It accounted for 7 of the 57 satellites from the region in early 2023. At the same time, Nigeria own the third largest number of satellites sponsored in the continent, and slightly surpassed by South Africa and Egypt, which as at then sponsored 12 and 13 satellites respectively (Oyewole 2020). Following the record of Egypt in 1998, South Africa in 1999, Morocco in 2001 and Algeria in 2002, Nigeria became the fifth African country to sponsor or co-sponsor a satellite with the orbiting of NigeriaSat-1 in 2003 (Oyewole 2017; Froehlich and Siebrits 2019).

As evident in Table 4.3, four of the seven satellite sponsored by Nigeria are for EO, while two are for communications. The most recent one is a military satellite, Delsat-1, of which there is limited public information. Two of the EO satellites were fully constructed by the UK based Surrey Satellite Technology Ltd (SSTL) with an agreement that entails technology transfer, which include training of Nigerian engineers. NigeriaSat-1 with 100 kg mass cost 13 million to manufacture in 2003, and was launched from Plesetsk Cosmodrome, Russia. Before NigeriaSat-1 went out of service in 2012, two other EO satellites were launched into orbit for its replacement in 2011. NigeriaSat-2 and NigeriaSat-X with 300 kg and 100 kg weight mass respectively were designed and manufactured at the SSTL, and launched from the Dombarovsky Air Base (DAB), Russia, at the total cost of US$ 112 million. As part of the technology transfer agreement with the SSTL, NigeriaSat-X was fully designed and built by Nigerian engineers, after the completion of their training, using its satellite AIT facilities. Due to the quality of its production and favourable orbital spot, the Nigerian EO satellites surpassed their operational lifespans (Oyewole 2021).

Table 4.3 Development of Nigeria's satellite inventories and capabilities

Satellite names	Mass of satellites (kg)	Manufacturer	Purposes	Launch and operational dates	Launch facility and locations	Cost (US$ million)
NigeriaSat-1	100	SSTL	Earth Observation	26/9/2003–2012	Plesetsk Cosmodrome, Russia	13
NigComSat-1	5,150	CGWIC	Communications	13/3/2007–11/11/2008	Xichang Satellite Launch Centre (XSLC), China	300
NigComSat-1R				17/12/2011		
NigeriaSat-2	300	SSTL	Earth Observation	17/8/2011	Dombarovsky Air Base (DAB), Russia	112
NigeriaSat-X	100	NASRDA engineers in SSTL lab				
EduSat-1	1	FUTA engineers in KIT		3/7/2017–13/5/2019	Kennedy Space Centre (KSC), USA	0.1
Delsat-1	–	–	Intelligence, surveillance and reconnaissance	4/2023	China	–

Source Compiled by the author with data from UCS (2020); Oyewole (2017); Gunter's Space Page (2020)

In 2004, the FGN contracted China Great Wall Industry Corporation Limited (CGWIC) for designing, manufacturing, launching, in-orbit services and insurance of a communication satellite as well as construction of its ground station and training of host engineers. It was the first of such broad services and technology transfer by China to any country. Launched in March 2007, the 5,150 kg mass NigComSat-1 that was designed to last for 15 years ran out of power due to technical anomaly, and eventually failed in orbit in November 2008. A new contract was signed by the two parties in early 2009, which led to the construction and successful launch of a replacement communication satellite NigComSat-1R in December 2011, making it the third satellite launched for Nigeria that year. It was revealed that the new satellite was replaced by the insurance cover with no additional cost to Nigeria. Hence, the total cost of these two communication satellites, which were both launched from the Xichang Satellite Launch Centre (XSLC), China, was US$300 million. However, these projects were largely funded by Chinese state-owned bank loan to fast-track their deliveries.

Starting from 2015/2016, the Japan government sponsored a space technology transfer initiative to the developing countries that is known as "Bird Project", which has been hosted by the Kyushu Institute of Technology (KIT). This project has benefited many countries, including Bangladesh, Ghana, Kenya, Nigeria, Mongolia, Thailand and Taiwan. Nigeria is one of the first set of countries to benefit from this initiative, which allows the engineers of the selected countries to develop capacity in satellite manufacturing under the supervision of KIT. Edusat-1, a research EO micro-satellite of 1 kg mass, is a product of this initiative. This satellite was built by Nigerian engineers from FUTA in the KIT AIT facility, and launched aboard SpaceX Falcon 9 from Kennedy Space Centre (KSC) in the USA. Edusat-1 was operated in constellation with other satellites under Bird project and tracked from Nigeria and other partner countries, before it was deorbited in May 2019. This project was cosponsored with a counterpart found of US$100,000 paid by FUTA with the support of the NASRDA (CESRA 2017). There are indications that a follow-up satellite Edusat-2 is currently is in production.

In the last five years, Nigerian government officials have severally announced plans and ongoing efforts to launch more satellites soon. In early 2016, the FGN announced its plan to obtain a US$600 million loan from China for the construction of NigComSat-2 and 3 (Oyewole 2017). An insider in the organisation acknowledged that the project of manufacturing these satellites is currently ongoing in China. Insider in the DSA equally acknowledged that the military was planning to have a research satellite produced and launched into orbit in 2020. Perhaps, this turn out to be the Delsat-1 that was launched in 2023 for military intelligence, surveillance and reconnaissance (ISR). For years, officials have also talked about ongoing NigeriaSat-3 project, which was initially schedule to be fully made in the country. However, one of the project insiders revealed the delay with the completion of Nigerian satellite AIT has changed the plan, which is now to produce part of the components required for the new satellite locally. Moreover, the Nigeria's 2021 appropriation bill shows that the NASRDA is sponsoring the production of two satellites. Amidst these, Nigeria tends to be underutilising available human resource capacities in local production of satellites.

Beside satellite inventory, Nigeria also have other interests in space-based capabilities, which are currently underdeveloped. The national space strategy roadmap entails the desire of the country to orbit a Nigerian scientist. This goal was initially set to be realised in 2015, but was later postponed to 2030, after the failure to meet the first targeted timeline. While some insiders are optimistic about the realisation of this desire within the new timeline, others are not. Currently, there are little or no indication to show that the country is preparing for this kind of mission, beyond a series of declarations by relevant government officials as well as enduring optimistic claims of some enthusiastic members of the scientific community. It is on this basis that the attention of this study will be turned to examine the development of space access, with the emphasis on rocketry and space launch programmes, in Nigeria.

4.4.2 Development of Space Access Capabilities in Nigeria: Space Launch Vehicles

Nigeria currently lacks space access capability or capacity to launch an object into the orbit. As evident in the previous section and Table 4.3, all the satellites sponsored by Nigeria were launched into orbit for the country from elsewhere. Nigeria is not alone in this case; no African country has the capacity to launch satellite into orbit currently (Gottschalk et al. 2018; Oyewole 2017, 2023a). Nigeria like other African countries have patronised Russia, China, and the USA among others for orbital launch services in the past. Some other African countries have equally patronised France and India for the same purpose. This trend is not likely to stop soon, especially for Nigeria. However, this does not mean that Nigeria does not have interest and commitment to developed required capacity to launch its own satellite from its own territory. In contrast, Nigeria is developing its space launch programme, which some government officials believed that it will be cost effective, given the location advantage of the country's closeness to the equator.

The CSTP is the institution that is principally charged with the responsibility to develop Nigeria's capacity in space transportation. The DSA equally has the responsibility to support national interests in this consideration. The national space strategic roadmap entails the desire of the FGN to launch a satellite into orbit from its territory by 2030. Very little is known about the progress that has been made in this consideration perhaps until recently, and many insiders avoided discussion about it. A report by Space in Africa shows that CSTP launched 30 experimental rockets, which were all below 10 km altitude, before 2019 (Ibe 2019a). As of 2020, however, an insider revealed that an experimental rocket that surpassed this altitude has been launched. The country has entered into partnership with some Asian and European countries to develop its capacity in this consideration. Beside the state sponsored rocket programme, several cases of amateur rocket flights that were sponsored by private individual, mostly rocket enthusiastic students, have also been reported in Nigeria over the years. Nevertheless, Nigeria has a long way to go in order to realise its objective with space access.

4.5 Conclusion

This chapter examines space institutions across public (governmental and non-governmental) and private realms, their mobilisation capacities in terms of financial, human and material (facilities and equipment) resources, and cumulative capabilities of the country in space assets and access. Beyond numerous centres of excellence, the Nigerian government creditably established African first dedicated agencies for civil and military programmes and fully-state owned corporation in space. However, the involvement of private sector and NGOs in Nigeria's space policy and programmes remain largely limited, given the fewer number of known actors in this area. Although Nigeria has equally committed significant financial resources and developed a growing human and material resources for its space programmes, their utilisations remain below optimum level. This is partially evident in the discussion on the capabilities of the country in space assets and access, and more importantly their contribution to socio-economic development and security. Accordingly, the focus of this chapter on space capacities and capabilities of Nigeria provides a bridge between the last chapter, which assessed the policy and strategic interests of the country in space, and the next chapter that examines the extent that such objectives have been realised.

References

Abiodun, A.A. 2008. Future trends in the global arena: Challenges and opportunities for Africa. *African Skies* 12: 12–22.

Adebajo, K. 2020. NigComSat: Nigeria's satellite company still not profitable after 14 years. *Space in Africa.*

AFL. 2020. Atlantic factorial limited. Available at https://atlanticfactorial.com/. Accessed on 8 Oct 2020.

Anderson, M. 2007. Nigeria space program isn't a 419 scam. *Wired.*

ARCSSTE-E. 2020. African regional centre for space science and technology education-English. https://arcsstee.org.ng/. Accessed 19 June 2020.

Argoun, M.B. 2012. Recent design and utilization trends of small satellites in developing countries. *Acta Astronautica* 71: 119–128.

ASTAL. 2020. Advanced space technology application laboratory. http://www.astaluyo.orgng/. Accessed on 17 June 2020.

AUST. 2020. African University of Science and Technology. https://aust.edu.ng/. Accessed on 17 June 2020.

AWB. 2020. AWB Nigeria. Available at https://awbnigeria.org/about/. Accessed on 7 Oct 2020.

BeepTool. 2020. About. Available at https://www.oyi.beeptool.com/inner-page/about-us/. Accessed on 8 Oct 2020.

Blueprint. 2020. Re: Allegation of corruption at NIGCOMSAT.

Bochinger, S. 2016. Space exploration in the space economy. Paper presented at Symposium on the Role of Industry in Space Exploration, United Nations, Vienna.

CAR-NASRDA. 2020. Centre for atmospheric Research. https://carnasrda.com/. Accessed 19 June 2020.

CAR-NASRDA. 2016. *Summary of activities for 2015 and 2016*. Ayingba: CAR-NASRDA.

CBSSA. 2020. Our historical background. https://www.nasrdacbss.com/our-historical-backgroud/. Accessed 19 June 2020.

CESRA. 2017. *Progress report on the activities of the centre for space research and applications (CESRA), [Volume 1 (Oct, 2014–July, 2017)]*. Submitted to The Vice Chancellor, The Federal University of Technology, Akure.

COPIN. 2018. Cooperative information network. http://www.copine.org.ng/. Accessed on 17 June 2020.

Daily Champion. 2011. Nigeria: FG orders NigComSat, others to move assets, personnel.

DSA. 2020. Defence space administration. https://dsa.mil.ng/. Accessed on 17 June 2020.

Elebeke, E. 2019. Breaking: Privatization of NigComSat will jeopardize national security-Pantami. *Vanguard.*

Euroconsultant. 2017. Government spending in space programs reaches $62 billion in 2016.

Fagbemiro, O. 2020. Founder's address, AWB Nigeria. Available at https://awbnigeria.org/about/founders-address/.

Fashade, O.A. 2008. National and regional capacity building activities for the acquisition of rocket launch capacity and the associated infrastructure–Nigerian strategy. *African Skies* 12: 86–89.

FGN. 2019–2020. *Appropriation Act, 2018–2020.*

FGN. 2020. *Appropriation Bill, 2021.*

FMST (Federal Ministry of Science and Technology, Nigeria). 2021. *National space policy and programme.* Abuja: Federal Government of Nigeria.

Francis, N. 2019. BPE to privatise 10, remaining NIPPs, NigComSat. *This Day.*

FRN. 2010. *National space research and development agency act, 2010.* Abuja: FRN.

FRN. 2016. *Defence space administration act, 2016.* Abuja: FRN.

Froehlich, A., and A. Siebrits. 2019. *Space supporting Africa: A primary needs approach and Africa's emerging space middle powers.* Cham: Springer.

FUTA. 2018. Centre for space research and applications. http://cesra.futa.edu.ng. Accessed 24 June 2020.

Gottschalk, K., P. Waswa, and S. Oyewole. 2018. The African space race. Filling Space. https://filling-space.com/2018/09/28/the-african-space-race/.

GSN. 2020. Geoinformation society of Nigeria. Available at https://geoson.org/about-2/. Accessed on 8 Oct 2020.

Gunter's Space Page. 2020. https://space.skyrocket.de/.

Harding, R.C. 2013. *Space policy in developing countries: The search for security and development on the final frontier*. New York: Routledge.

Ibe, J. 2019a. Towards Africa's first rocket launch from its soil—NASRDA Centre for Space Transport and Propulsion is making progress. *Space in Africa.*

Ibe, J. 2019b. How much has NASRDA contributed to Nigeria's economic growth and development. *Space in Africa.*

ISSE. 2019. Institute of space science and engineering. https://isse.edu.ng/.

Lal, B., E. J. Sylak-Glassman, M. C. Mineiro, N. Gupta, L. M. Pratt, A. R. Azari. 2015. *Global trends in space volume 1: Background and overall findings.* IDA Paper P-5242, Vol. 1 (June). Washington, DC: IDA Science & Technology Policy Institute.

Luncedo, N., and K. Gottschalk. 2013. The growth of space science in African countries for Earth observation in the 21st century. *South African Journal of Science* 109 (1–2): 1–5.

Martinez, P. 2009. Building capacity in the basic space sciences in Southern Africa: Experiences and prospects. *Advances in Space Research* 43: 1866–1872.

Martinez, P. 2012b. Is there a need for an African space agency? *Space Policy* 28: 142–145.

Martinez, P. 2016. The development and initial implementation of South Africa's national space policy. *Space Policy* 37: 30–38.

Martinez, P. 2012a. The African leadership conference on space science and technology for sustainable development. *Space Policy* 28: 33–37.

Mistry, D., and B. Gopalaswamy. 2012. Ballistic missiles and space launcher vehicles in regional powers. *Astropolitics* 10 (2): 126–151.

Mohammed, S.O. 2011. *Nigerian space research and development agency. Director-general presentation.* Abuja: NASRDA.

Munsami, V., and E. O. Offiong. 2017. Should Africa follow the European space governance model for an African space programme? *Space Policy* 30: 1–6.

NASRDA. 2020. NASRDA centers. https://nasrda.gov.ng/en/about-us/nasrda-centers/. Accessed on 15 June 2020.

NBS. 2017. *Annual abstract of statistics 2016*. Abuja: Nigerian National Bureau of Statistics (NBS).

NCRS. 2019. National centre for remote sensing. http://www.ncrsjos.org/. Accessed on 17 June 2020.

NMETS. 2011. International conference of the Nigerian meteorological society (NMetS) 'Zaria 2011'.

Noe, R.A., J.R. Hollenbeck, B. Gerhart, and P.M. Wright. 2011. *Fundamentals of human resource management*, 4th ed. New York: McGraw-Hill.

NSE. 2020. Nigerian society of engineers. Available at https://nse.org.ng/about-nse/. Accessed on 9 Oct 2020.

NSPS. 2020. Nigerian society of physical sciences. Available at https://nsps.org. Accessed on 9 Oct 2020.
Oludimu, T. 2018. Why we known next to nothing about Nigeria's space programme. *Techpoint Africa.*
Osuagwu, P. 2012. Battle over NigCOmSat bill shift to senate. *Vanguard.*
Othman, M. 2003. Small satellites for the benefit of developing countries. *Acta Astronautica* 52: 687–690.
Oyewole, S. 2017. Space research and development in Africa. *Astropolitics* 15 (2): 185–208.
Oyewole, S. 2020. The quest for space capabilities and military security in Africa. *South African Journal of International Affairs* 27 (2): 147–172.
Oyewole, S. 2023a. The African orbital journey in space age: From spectator to participant. *Vanguardia Dossier* 88 (15): 72–75.
Oyewole, S. 2021. One of Nigeria's satellites is on its last legs: why this is worrying. *The Conversation.* https://theconversation.com/one-of-nigerias-satellites-is-onits-last-legs-why-this-is-worrying-165068.
Oyewole, S. 2022. Air and space strategy in African counterinsurgency operations: Insights from the Nigeria's War against Boko Haram. Presentation at the International Conference on Counter-terrorism, Technology and Development in Africa, jointly organised by the Centre for Military Studies, Stellenbosch University, South Africa and the Africa Research Institute, Óbuda University, Hungary.
Oyewole, S. 2023b. Military space strategies and African realism: Egypt, South Africa and Nigeria. In *Oxford handbook of space security*, ed. S. M. Pekkanen and P. J. Blount. Oxford: Oxford University Press.
Pen Pushers. 2020. NASRDA: Nigeria's latest home of corruption.
Punch. 2016. NigComSat-1R becoming white elephant four years after—investigation.
Sahara Reporters. 2020. Group writes petition against NIGCOMSAT managing director over gross financial misappropriation.
Seminari, S. 2019. Global government space budgets continues multiyear rebound. *Space News.*
Senyucel, Z. 2009. *Managing human resource in the 21st century*. Telluride, CO: Ventus Publishing Aps.
SGAC. 2020a. About us. https://spacegeneration.org/about. Accessed on 21 June 2020.
SGAC. 2020b. SGAC. Available at https://spacegeneration.org/. Accessed on 21 June 2020.
Shabbir, Z., and A. Sarosh. 2018. Counterspace operations and nascent space power. *Astropolitics* 16 (2): 119–140.
Sims, R.R. 2002. *Organizational success through effective human resources management*. Westport: Quorum Books.
Space in Africa. 2018a. South Africa ad Nigeria supremacy on space investment.
Space in Africa. 2018b. 8th national conference of the astronomical society of Nigeria-Nov 21–24.
Space in Africa. 2019a. *African space industry annual report* (2019 Edition). Lagos.
Space in Africa. 2019b. *NewSpace Africa industry report*. Space in Africa.
Space in Africa. 2020a. The future of the Nigerian space agency and the Nigerian communications satellite limited.
Space in Africa. 2020b. Nigeria reps to carry out forensic audit on NigComSat.
The Nation. 2018. Workers threaten to shut down NIGCOMSAT over alleged victimisation.
The Whistler. 2019. Investigation: How Nigeria's space agency officials looted over N300 million (Part 1 &2).
TheNews. 2019. Climate change now a critical threat in Nigeria—experts at FUTA conference.
UCS (Union of Concerned Scientists). 2020. Satellite Database. Available at https://www.ucsusa.org/resources/satellite-database. Accessed on 28 Jan 2020.
UN-CPUOS (United Nations Committee on The Peaceful Uses of Outer Space). 2006. Report of the Scientific and Technical Subcommittee on its forty-third session, agenda item 8, forty-ninth session, Vienna, 7–16 June 2006.
UNN. 2020. Centre for basic space research. http://www.unn.edu.ng/academics/centres/centre-for-basic-space-research. Accessed 6 June 2020.

van Wyk, J.A. 2009. South Africa's space policy and interests: A new dawn or a black hole? *Strategic Review for Southern Africa* 31 (2): 46–73.
Wood, D., and A. Weigel. 2011. Building technological capability within satellite programs in developing countries. *Acta Astronautica* 69: 1110–1122.
Wood, D., and A. Weigel. 2012. Charting the evolution of satellite programs in developing countries—the space technology ladder. *Space Policy* 28: 15–24.
Wood, D., and A. Weigel. 2014. Architectures of small satellite programs in developing countries. *Acta Astronautica* 97 (2014): 109–121.

The Contribution of Space Policy to Development and Security in Nigeria

5

Abstract

This chapter examines the extent that space policy and strategy have advanced the development and security interests of the government and people of Nigeria. It interrogates the convergence and divergence of space capacities and capabilities (such as institutional, financial, human and material resources, access and assets), and the realisation of set policy objectives. It explores records and gaps in space support for national competitiveness, economic growth and transformation, local technological capacity, defence, security, power projection, disaster management, human rights and the struggle for freedom, democracy, effective and good governance, food and water security, sustainable environment, quality of health and education systems, and international cooperation as indicators for national and human dimensions of development and security.

5.1 Introduction

The development of Nigeria's space policy and programmes is evident with the growing profile, capacities and capabilities of the country in the sector. As shown in chapters three and four, the development of space policy, strategy, institutions, financial, human and material resources, and relevant space assets offers a sectorial perspective to national development in Nigeria. However, this sectorial development further raises questions on the bigger picture of national and human development and security in the country. The big question is: how has the development in the space sector advanced the broader development and security aspirations of the people and government of Nigeria? One can also quarry why the government has decided to prioritise space in institutional building, human resource and infrastructure developments, and investment, in the face of pressing national and human development ad security challenges.

S. Oyewole, *Utilitarianism in Outer Space*, Southern Space Studies,
https://doi.org/10.1007/978-3-031-49646-2_5

This chapter examines the extent that space capacities and capabilities have achieved the desire policy and strategic objectives, especially the development and security aspirations of the people and government of Nigeria. In doing this, this chapter pays attention to both the state-centric and human/people centric interests and perspectives on development and security vis-à-vis space policy and strategy. For this purpose, this chapter is further divided into three sections. The first section deals with state-centric perspective, as it examines space support for national development and security in Nigeria. The second section of this chapter focuses on people-centric perspective on space support for sustainable human development and security in the country. The final section provides a concluding thought on the subject matter of the chapter.

5.2 Space Support for National Development and Security in Nigeria

From a state-centric perspective, the Nigeria's space policy and strategy were designed to promote and support the aspiration of the government of the country. This section is interested in the extent that this objective has been achieved. For this purpose, this section is further divided into two: the first subsection focuses on space support for national development, and the other subsection examines space support for national security, in Nigeria.

5.2.1 Space Support for National Development in Nigeria

As detailed in chapter two, this book identifies different conceptions and perspectives on development, of which some of them focus on national development. These ends include economic growth and transformation, resource management, competitiveness and catching up with others, local or independent industrial capacity among other things that are considered as objectives of development. These among others reflect some aspects of national space policy objectives and development conceptions of the government of Nigeria, as reviewed in the chapter three of this book. Some of these objectives will be considered in relations to space policy, strategy, capacities and capabilities of Nigeria as follow:

5.2.1.1 Development as Competitiveness and Catching Up

International competitiveness, catching up with advanced states, or staying comparable with peers constitute a dimension of national development goal. Nigeria is generally recognised as a developing country and emerging economy and power globally, and leading power with the biggest economy regionally (Adebajo 2017; Amusan and Oyewole 2017; Ogunnubi and Okeke-Uzodike 2016; Ogunnubi and Oyewole 2020). From a global point of view, the policy and strategic interests of Nigeria, like many other developing or emerging countries, in space include catching up or closing the development gap with the developed countries. In this consideration, the country has prioritised the need to acquire essential capabilities, including human and technical capacities, to derive benefits from space and attendant technologies. This is responsible for the growing capacity of the country to access space products and convert them to its consumable need. Nevertheless, Nigeria largely remains a consuming nation in term of space technologies, with little or no facilities, assets, inventions, productions, and market share of a major space player globally.

Among the developing countries and emerging powers or players in space, or regional powers globally, Nigeria is interested in staying comparable, catching up with those ahead and staying ahead of those behind. This is equally connected to Nigeria's regional leadership status and aspiration in the continent of Africa. Many stakeholders in Nigeria's space sector see no reason why the country should not lead the continent and stay competitive among developing countries. Nigeria's policy aspirations and public institutionalism in space can be better rated than many other emerging space players globally. However, the country remains an average player among emerging space powers, with little industrial capacity to support its space programmes. Compared to other regional powers outside Africa or emerging space powers like India, Brazil, Pakistan, Israel, Argentina, Indonesia, Iran, United Arab Emirates (UAE), South Korea and North Korea, it is apparent that Nigeria fall behind in term of space access, facilities, assets, inventions, productions, and market share (cf. Al-Ekabi and Ferretti 2018; Al-Ekabi et al. 2017; Burger and Bordacchini 2019; Harding 2013; Harvey et al. 2010; Lal et al. 2015; Lopez 2015; Mistry and Gopalaswamy 2012; OECD 2014; Space Foundation 2015).

In the continent of Africa, Nigeria is recognised as one of the most advanced space players, which is only comparable to South Africa, Egypt, Algeria and perhaps Morocco. As of late 2020, these are the only countries that have sponsored more than a satellite as well as have owned and operated them for about two decades or more (Froehlich and Siebrits 2019; Luncedo and Gottschalk 2013; Oyewole 2017a, 2020a, 2023a). Amidst these, Nigeria is leading in terms of space policy aspirations and public institutionalism. It was the first African country to establish a space agency in 1999 and publish its national space policy and strategy in 2001. Algerian Space Agency was established in 2002, SANSA came into existence in December 2010, Egyptian Space Agency was established in August 2019, while Morocco is yet to create an agency for this purpose—its space programme is largely coordinated by a centre for remote sensing. Since 2015, Kenya, Ethiopia, Zimbabwe, Rwanda, and Uganda among others have equally joined the list of few African countries that have advanced the institutionalism of their space programmes from a mere centre of remote sensing to an agency status. Equally, Nigeria established the African first fully state-owned corporation for space commercialisation in 2003 and agency for military space programme in 2014.

Despite the foregoing signs of competitiveness, the Nigeria's space programme also started with the desire to catch up with other regional powers in the continent. The launch of the first orbited satellite from Africa in 1998 was sponsored by Egypt, which was followed by South African satellite in 1998, Morocco and

Germany co-sponsored satellite in 2001, and Algerian satellite in 2002, before the Nigerian own was launched in 2003. This considerably informed the Nigeria's ambitious space policy and programmes between 1999 and 2007, including the desire and launch of the first African state sponsored communication satellite. It is also important to note that Nigeria's defence space programme started in 2014, the same year that African first military satellite sponsored by South Africa was launched. South Africa and Egypt are equally ahead of Nigeria in astronomical observatories and some other facilities for basic space sciences, capacity building for space access, and commercialisation of space products. There are indications that South Africa, Egypt, and recently Algeria and Tunisia have functional AIT for designing and manufacturing satellite, while Nigeria's facility is yet to be operational.

Egypt and South Africa are further ahead of Nigeria in total number of satellites sponsored. As of late 2020, Nigeria had only sponsored six satellites, while Egypt had sponsored nine satellites and South Africa eight satellites. In early 2023, Nigeria had sponsored seven satellites, while Egypt and South Africa had sponsored 13 and 12 satellites, respectively. However, Nigeria may catch up with these records with planned two EO satellites, two communication satellites and more satellites for technological demonstration. Yet, it is also important to note that Nigerian satellites mostly have longer lifespan, which offer the country more enduring services, when compare with South Africa. Generally, Nigeria remains one of the leading African space players, although this status is increasingly challenged with the emergence of other countries, which are also struggling to catch up and close the development gap in the region.

5.2.1.2 Development as Economic Growth and Transformation

Nigeria space programming has proved relevant to national development, as it concerns economic growth and transformation. Space has been employed to transformed different sector of the economy. However, the extent of the impact of space programme has not received adequate attention in most cases. Prominently, the space programme has supported communication, digitalisation, urban planning, construction, resources management, and education, which have all boosted the economy and created wealth for many, in Nigeria.

Nigeria has recorded progress in adopting communication technologies and digitalisation of its economy in the last two decades. This is considerably evident with the rapid adoption and access to mobile phone and internet penetration, which has greatly revolutionised the country. Mobile cellular subscription in Nigeria rose from 0.024 per 100 people in 2000 to 88.18 per 100 people in 2018. In 2000, only 0.06% of the population had access to internet services in Nigeria. In 2017, however, 42% of the country's population had access to internet (World Bank 2020). This development, which is a product of a series of policy choices in relation to terrestrial and space infrastructures, has revolutionised different sectors of the economy, including agriculture, manufacturing, and services. Most notably, the banking sector has been revolutionised with the growing adoption of e-banking and mobile banking. The growing access to mobile phone and internet has spinoff e-commerce, e-learning, e-governance, precision agriculture, and telemedicine. In this case, satellites are playing some considerable roles, most especially to connect the unconnected and the under-connected peoples in the remote parts of the country. Nevertheless, the NCC recently revealed that satellite only contributes as low as 0.2% to Nigeria's internet connectivity (Akinyemi 2020).

In addition to the trend of internet access and its revolutionary impact in Nigeria, the huge and growing access and subscription to satellite TV and radio in the country constitute another aspect of space support for economic growth and transformation in the country. These platforms are transforming access to information, innovation, and e-commerce. Prominent among the impacts of these platforms is the transformation of Nigeria's entertainment industry, which now worth billions of naira. However, there are indications that most satellite TV and radio service providers in Nigeria preferred foreign data, a trend that has undermined the profitability of NigComSat. With the launch of direct to home platform by NigComSat Ltd, most private firms that are involved in satellite TV platform services considered the corporation as a competitor rather than collaborator. Consequently, they preferred alternative sources of satellite data, and have supported the enduring campaigns for NigComSat privatisation.

Through more availability and accessibility to remote sensing data from satellite, Nigeria's space policy and programmes have transformed and supported construction, agriculture, resource management and other sectors of the economy. Prior to this development, it was gathered that visibility report for federal road construction in two decades ago usually takes months or years, but similar exercises are carried out and completed in days or weeks now due to space support. In essence, space support provides fast and reliable mapping of terrains in manner that are very helpful to assess the visibility and monitor the progress during construction of roads, rails, bridges, dams, ports, and other notable infrastructure projects in Nigeria. But while the impact of space assets is revolutionary for this sector, it does not eliminate some of the societal ills that are associated with public infrastructural developments in the country. These include corruption in pricing, politicisation of biding and contracting, inadequate quality or failure in delivery, and other questionable human factors or dealings in the sector.

The impacts of space support are equally noticeable in the agriculture sector. Over the years, the space agency has provided the Ministry of Agriculture and associated departments and agencies with satellite data that are relevant to policy planning and assessment in Nigeria. Satellite data that are made available to universities and research centres across the country have also boosted agricultural research and development in the country. The space policy and programme have equally encouraged or sustained the rise and activities of some consulting firms and satellite service providers that are providing supporting services to farmers. However, these services have remained largely limited to few large-scale farmers, while they are mostly less accessible and unavoidable for majority of Nigerians that are involved in this sector.

The space programme has supported Nigeria's struggle to improve its capacity in natural resources management. Nigerian satellites have helped the country to compile inventories of its natural resources, mining sites and its environmental impacts. The FGN mandated the NASRDA to develop resource inventory of the country in 2007. In response, the agency used NigeriaSat-1 to produce several high-resolution satellite images that worth 4 billion Naira (US$11 million) in its first phase of resource mapping of the country. Following the launch of NigeriaSat-2 and X in 2011, the agency commenced the second phase of the survey, which is conducted at the scale of 1: 50,000 in most cases. This project was estimated to be worth 5 billion Naira (about US$13.8 million) upon completion (Space in Africa 2019b). These are important in a resource dependent economy like Nigeria, a country that the total natural resources rent is 15.08% of the GDP on the average between 1970 and 2017 (World Bank 2020). Moreover, the efforts of space programming for environmental impacts assessment in mining sector of the country will receive better attention later in this chapter.

Despite many worrisome trends, especially in relation to inadequate utilisation of most of the available assets, the space policy and programmes have made significant impacts and contribution to the economic growth and transformation of Nigeria. As it will be shown later, the contribution of the space programming to the quality of education and health in Nigeria is financially significant. However, the revenues generated by space assets, firms and agency offers more insight into the contribution of this sector to economic growth in the country. In total, the NigeriaSat-1 directly contributed over 10.5 billion Naira (about US$29 million) to Nigeria's economy within its first 9 years in orbit (Space in Africa 2019b). Accordingly, this satellite considerably appears to be profitable, given that the total cost of manufacturing it is US$13 million.

The net worth of space companies in Nigeria can also reflect their contribution to the economy. The annual revenue of the Lagos-based Atlantic Factorial with 15 employees was estimated between US$500,000 and US$1,000,000 (Space in Africa 2019b). With 13 employees, Beeptool raised US$600,000 investment in 2019. The company was recently evaluated to worth US$12.5 million, even with less than US$100,000 annual revenues. Equally, GeoApps was reportedly declared by its leadership to worth US$1 million (Space in Africa 2019b). While NigComSat has not disclosed it revenue publicly, the corporation is believed to be one of the biggest corporations in African space industry. These among others show the contribution of space programming to economic growth and transformation in Nigeria.

5.2.1.3 Development as Local/Independent Capacity and Self-reliance

The Nigeria's space policy strategies echoed some of the objectives of development that are known to post-colonial perspective on utility (Akinrinde and Oyewole 2021; Akinyede and Boroffice 2013; Froehlich and Siebrits 2019). Prominent among these is the need to develop local capacity and self-reliance in different aspects of space capabilities. As earlier discussed, Nigeria has acquired some space technologies and developed institutional capacity to covert those resources into utility, for the advancement of its development and security. Prior to the advent of coherent space institution, policy, strategy, programme, capacities and capabilities between 1990 and 2000s, Nigeria was a mere consumer of space products. Hence, the country lacked the capacity for direct access to space products and covert them as required for its need.

With the launch of Nigerian satellites and development of their ground stations, the country has maintained capacity for direct and unrestricted access to various satellite services and utilised them as deem suitable for its national interests. The country has equally acquired considerable capacity in satellite design and manufacturing, although inability to complete its AIT facility has undermined its potentials, considering the quality and quantity of human resources that are available to the country in terms of space R&D. Despite the progress made by the country in rocketry as well, it quests for independent orbital launch capabilities remain a long-term goal. However, the progress made in space assets acquisition and their utilisation, policy articulation and operationalisation, institutional development, resource mobilisation and conversion are significant indicators of increasingly local and independent capacities developed in Nigeria.

5.2.2 Space Support for National Security in Nigeria

Space capabilities, most especially the use of satellites for communication, remote sensing and navigation have been employed to advance Nigeria's national security interests. These include in the area of defence, power projection, law enforcement and disaster management. Nigeria's space capabilities have contributed to the efforts of the country to neutralise or minimise internal and transnational threats against the country. These include threats of terrorism, insurgency, banditry, piracy, smuggling, kidnapping, oil theft and bunkering, illegal mining and refining, and many other forms of political or criminal violence. These capabilities have further contributed to Nigeria's peacekeeping, peace enforcement and military power projection across Africa and beyond. Space support has also featured in Nigeria's strategic planning and management of natural and man-made disasters.

5.2.2.1 Space Support for Internal Security Operations

Boko Haram and Nigeria's Counterinsurgency Operations

One of the greatest security threats that has confronted Nigeria in the twenty-first century is the rise of Islamist Boko Haram militancy and insurgency (Oyewole 2013). Boko Haram emerged from northern Nigeria and expanded into neighbouring countries in the Lake Chad region, including Cameroon, Chad and Niger Republic. The group developed network with Al Qaeda and its affiliates, such as Al Qaeda in the Islamic Maghreb (AQIM) and Al Shabab, as well as the Islamic State in Iraq and Syria (ISIS) and its affiliates in Sahel and North Africa (Oyewole 2018a). Boko Haram insurgency and terrorism are characterised by (suicide) bombing, assaults, hostage taking and ransoming, assassination, sexual violence, raids against border communities, military and police targets, as well as land grab. At the peak of its campaign of violence between 2014 and 2015, its membership strength ranges from 15,000 to 50,000 (Oriola et al. 2021). The group has directly or indirectly killed more than 300,000 children, kidnapped tens of thousands, and displaced over 5 million people across the region (Oyewole 2023d). The group controlled 20,000 square miles of Nigeria's territory at its peak between 2014 and 2015 (Onuoha and Oyewole 2018).

The Nigerian government has responded to the threats of Boko Haram through economic, military, diplomatic, political and socio-cultural engagements. Amidst these, the armed forces have been deployed with the law enforcement agencies (police, customs, immigration, Department of State Service (DSS), the Nigeria Security and Civil Defence Corps (SCDC), and the Civilian Joint Task Force (CJTF) among others), as a Joint Task Force (JTF) to contain Boko Haram insurgency, as well as restore stability and improve security in the affected region. In this arrangement, the army and the air force have played prominent roles in the campaign against the insurgents in Nigeria and the Lake Chad region (Omeni 2018). The NA deployed two of its divisional strength, which is estimated between 40,000 and 50,000 personnel, to the theatre of operation, while the NAF equally maintains one of its six Command in the region (Oyewole 2017b, 2021b). Given the strategic location of Lake Chad, the NN is also involved in the operational theatre. Nevertheless, the critical role of the space institutions, such as the NASRDA, NIGCOMSAT and DSA, and their capabilities have received inadequate attention.

The counter-insurgency (COIN) operation of the Nigerian military in the Lake Chad region has been supported by space assets. For instance, when targeted attacks by the insurgents crippled the tele-communication facilities in the region, the military and other security forces in the region relied on satellite supported communication provided mainly by the NIGCOMSAT with little or no charges (Oyewole 2022a, 2023b). The NASRDA and DSA also supported the military with comprehensive survey, mapping, geospatial analysis of the operational theatres, and training of personnel in related matters. These are important in tactical and operational engagements of the military against the insurgents. Although Nigeriasat-2 & X are not designed to serve military purposes, their data are equally relevant in search and rescue of Boko Haram hostages. For instance, when the Chibok schoolgirls were abducted by Boko Haram in April 2014, the military was provided with detailed satellite data of the region to support planning of search and rescue mission. Although the hostages were found, the military decided against rescuing them, considering the suicidal threat made by the insurgents to the lives of their victims (Oyewole 2016c). However, thousands of other hostages were rescued by the military in many general sweep operations against the insurgents between 2015 and 2016 (Oyewole 2016b).

Satellite data have been relevant to the air and surface (ground and maritime) forces in their search and destroy operations against insurgent camps and logistics in the fast area covered by Sambisa Forest Reserve and other forests as well as fridge islands across the Lake Chad region (Oyewole 2022a). In addition to air and human intelligence, satellite data are relevant in mapping and planning general sweep operation to recover lost territory from the insurgents. When insurgent use of anti-air missiles tends to undermine air power capabilities and delivery of the NAF in 2014, the service and the space agency collaborated in R&D of satellite guided missiles, which enable the aircraft to fly at higher attitude with better precision strike (Oyewole 2020a). In this manner, Institute of International Security Studies (IISS 2020) recognised that Nigeria is the only Sub-Saharan African country with an air force that maintains Inertial Navigation System (INS)/GPS guidance capabilities for aerial bombing. This development is acknowledged to have increase the air strike capability and efficacy of the NAF (Oyewole 2021a).

Other Internal Security Operations in Nigeria

Beyond Boko Haram, space support is also evident in other internal security operations of the Nigerian military. The military has been deployed in support of the law enforcement agencies against subversive elements across Nigeria. This has been the case in the Niger Delta, where troops have been deployed against militancy, sabotage, kidnapping, ransoming, oil theft and illegal refineries that are designed to cripple the country's oil and gas sector. Military operations have been sustained in the region since 2003. As of 2016/2017, the army maintained a divisional HQ and nine brigade strength with estimated 13,500 to 31,500 personnel in the Niger Delta, which also hosted two of the three naval commands and ten operational bases with estimated 5,750–8,000 personnel as well as an air force command with six operational bases (Oyewole 2018b, 2020c). Equally, military deployment and operations against armed banditry are salient in the northern Nigeria, especially in Zamfara, Benue, Katsina, Sokoto, Taraba, Niger, Adamawa, Nasarawa, Plateau and Kano States. Since 2018, no fewer than 1,000 troops have been deployed to most of these states to combat armed bandits (Aina et al. 2023; Oyewole et al. 2022).

The NN has also played a leading role in the war against maritime piracy and robbery, illegal, unreported and undocumented fishing, among other crimes in Nigeria's internal waters and the Gulf of Guinea (Carnimeo and Oyewole 2020; Oyewole 2016a, 2023c). It has collaborated with other services of the armed forces and the paramilitary in the campaigns against on-shore infrastructures and sanctuaries of the Nigerian based pirates of the Gulf of Guinea (Daxecker and Prins 2021; Oyewole 2015a). Equally, the military has joined other security agencies in combating trans-border crimes, including smuggling of weapons and other barned products to and from the country (Oshita et al. 2019). The military is also combatting different forms of kidnapping across Nigeria, through a series of missions for search and rescue of hostages, and search and destroy of kidnapper's den (Omotola and Oyewole 2024; Oyewole 2016b).

In many cases, space-based survey and mapping have helped the military forces in planning and familiarising themselves with the operational terrain. In addition to human and air assets, satellite data have equally supported the troops in the efforts to locate and neutralise bandits, their camps and logistics. Military operation against militants, illegal miners and refiners, oil bunkers and saboteurs, armed robbers against ships and communities, kidnappers and pirates in the Niger Delta and other coastal communities in Nigeria have sometimes benefited from space-based C4ISR. In addition to coastal radars, the Nigeria's war against maritime piracy has equally benefited from space supported domain mapping and situation awareness, including in ship tracking, and emergency response to distress calls, especially through communication and navigation. Besides, the Nigeria's war against smuggling also benefitted immensely from space capabilities. Between August 2019 and December 2022, when the federal government of Nigeria ordered the closure of boarder with Benin Republic and established a military-led JTF to combat smuggling, the security forces at the frontline across the remote border communities and their command centres relied on satellite for their communication, command and control.

Despite the foregoing, there are observable underutilisation of space capabilities in Nigeria's security governance. Notably, the inadequacies of the country in town planning by some local government authorities, as well as forest and waterways management bodies have questioned the levels of space technological access and utilisation of relevant institutions. These have turned many forests, highlands, waterways as well as unregistered and uncompleted buildings across the country into sanctuaries for armed bandits, robbers, kidnappers, militants, terrorists, insurgents, pirates, smugglers, oil bunkers, and illegal mineral resources miners and refinery operators.

5.2.2.2 Space Support for External Security Operations

The Nigerian military has considerably enjoyed space support in its external engagements. As one of the leading military powers in Africa, Nigeria has been involved in a series of peacekeeping and peace enforcement operations and power projections in the continent. Between 1990 and 2018, Nigeria contributed 741,398 troops to various peace-keeping missions globally, which represents 12.2% of such contribution from African nations (Oyewole 2020b). Some of these missions have benefited from space support. Even before the nation's space policy and programme took a coherent shape, the Nigerian military-led ECOWAS Monitoring Group (ECOMOG) and UN supported operations in Liberia and Sierra Leone in the 1990s benefited from space supported command, control and communication. Satellite phone was critical for the military commands in Monrovia and Free Town to maintain communications with the authority in Lagos as well as link between troops and their families (Adesina 2002; Oyewole 2023b).

Over the last two decades, the Nigerian space establishments have played more active roles to support its military operations beyond the territorial confine of the country. Notably, the space agency supplies the military forces with relevant satellite imageries to support their operations in Mali, Sudan and Central African Republic. In 2012, Nigeria and ECOMOG allies benefited from space support in their peace keeping operations in Mali, especially in satellite mapping of the theatre for pre-deployment training, tactical management and operational planning. The space agency also provided the NAF with topographic map to support investigation into the Nigerian fighter jet that crashed in active operation against the insurgents in Niger Republic in 2013 (Space in Africa 2019a). The country also brought space support to bear for its military deployment to AU and UN missions in Sudan. These have contributed to Nigeria's capabilities to project military power and defend its interests, allies, commitments and status internationally.

5.2.2.3 Space Support for Situational Security Operations

Nigeria, like any other country across the word, is not immune to natural, technological and other complex disasters. Flood, storm, earthquake or tremor, landslide, drought, epidemic, wildfire, transport and industrial accidents, among others are not strange in different parts of Nigeria. Between January 1900 and July 2022, Nigeria accounted for 500 of the total 5,337 and 20,257 cases of disaster recorded in Africa and globally, respectively. Amidst these, the country recorded no fewer than 48,075 fatalities (Oyewole 2022b). Moreover, COVID-19 and associated lockdown also show how challenging it is to maintain law and order during situational crisis in Nigeria and beyond (Oyewole and Olajubu 2020). In these and other situational security challenges, disaster management strategy has considerably benefited from space support. This is notable in environmental crisis management, including natural and man-made disasters.

The Nigerian satellites have been instrumental in mapping dams, rivers and other water bodies to provide early warming and real time situation awareness to forecast and prepare for floods in the country and beyond. They have also provided early warning for tremor, landslides, earth-quack, volcanic eruption and bush fire across the world. Nigeria provided satellite data to support the South-East Asia in response to tsunami, USA in its response to hurricane Katrina, Brazilian response to wide fire in Amazon Forest, and response to locust invasion in Algeria. Cameroon, Chad, Niger, DR Congo, Australia, and Sudan among other countries have also benefited from Nigeria's space support in their responses to various natural disasters. In these and other ways, space capabilities have contributed to Nigerian security interests home and abroad. It is important to note, however, that space support for disaster management is not a sole determinant or guarantee of effective response of state and other relevant institutions.

Although space supported early warning and real time situation awareness or information are necessary to manage and possibly address threat of natural and other forms of disasters, they are not enough to guaranty effective strategic and operational responses. In many cases, early warning has received inadequate political will, policy attention, institutional preparedness, resource mobilisation, strategic management and public awareness to address or confront such impending dooms, and minimise its consequences. Despite the enormous amount of satellite data that are available and relevant to disaster management, weak institutional capacities of the state have continued to undermine and compromise Nigeria's situation security goals (Oyewole 2021a).

5.3 Space Support for Sustainable Human Development and Security in Nigeria

Beyond the traditional understanding and state-centric conception of development and security, space has supported human development and security in Nigeria. The development of space capacities and capabilities, and their contribution to national development and security have advanced human development and security goals. This is evident in ways that space has considerably supported job opportunities, wealth creation, poverty alleviation, education, competitiveness of the nation and dignity of its people, and disaster management. However, subjects of human development and security deserve more attention in this consideration. Accordingly, this section examines space support for human rights, good governance, food, water, and environmental security, quality of health and education systems, poverty alleviation, employment opportunities, and youth empowerment, as well as international cooperation.

5.3.1 Human Rights and the Struggle for Human Freedom

Nigeria's policy, which include programmes run with indigenous and foreign space capabilities, have supported the quest for human rights and good governance in the country. Satellite TV and radios as well as (terrestrial and space supported) internets have greatly revolutionised public space in Nigeria. These platforms have enhanced public access to information, and empowered the people to mobilise for better political representativeness in governance. Satellite radio and internet platforms played important role in the mobilisation of the Igbos for better political education, representation, integration and threat of secession, especially between 2014 and 2017 (Oyewole 2019). Prominent among the actors in this mobilisation are the Movement for the Actualisation of the Sovereign State of Biafra (MASSOB) and the Indigenous People of Biafra (IPOB). Besides, the #BringBackOurGirls campaigns between 2014 and 2018, and the #EndSARS campaigns between 2017 and 2020 relied heavily on internet platform to mobilise public support for the struggle for human freedom in the face of state power in Nigeria.

A growing number of media outlets and civil society groups in Nigeria are embracing the use of satellite images to prove cases of human rights violation against state and non-state actors, most especially against the military in the less accessible North East, where Boko Haram's campaign of terror have endured for more than a decade. Satellite data contributed to the available public information and the attendant political campaign against the underplayed fact of the accidental aerial bombing of Rann-based IDP camp, where 234 people were killed in January 2017. Similar data was equally useful in assessing the NAF bombing of communities, which were thought to be insurgent camps, in Denmsa and Numan LGAs of Adamawa State in December 2017 (Amnesty International 2018a, b;

Oyewole 2021b). Accordingly, satellite data has become a weapon that is available to media and civil society groups to uncover human rights abuses beyond state disinformation and area denial in conflict theatres in Nigeria.

Satellite data is further relevant to determine human rights abuses by non-state actors and the effectiveness of state responses in inaccessible areas in Nigeria. One of the bloodiest attacks ever recorded by Boko Haram was in Baga on January 3, 2015, where over 2,000 people were killed, 3,700 buildings were destroyed, about 15,000 people were forced to become refugees in Chad and many more were displaced internally within Borno State in Nigeria (Oyewole 2015b). However, the government and the military downplayed the attack and the impacts on the affected communities. Local and international media and NGOs were able to uncover the reality and raise public attention for the plight of the affected communities using a combination human and space assets. Satellite data of the destroyed communities were widely published. These among other publicity of security failure and attendant public campaign for transparency in security crisis handling affected the then siting government, making Goodluck Jonathan the first ever incumbent President of Nigeria to lost at the poll in the 2015 elections.

Despite the foregoing, the inadequate capacity of most Nigerian media and civil society groups in resorting, accessing, consuming and analysing satellite data and mobilising public support for convincing course have continued to undermine the utility of space support for human rights and associated struggle in the country. The inadequate connectivity, especially internet penetration, and literacy rate of the country have also limited the potential impacts of space supported messages and mobilisations of the media and civil society groups in Nigeria.

5.3.2 Democratisation, Policy Efficacy, and Good Governance

Beyond human rights campaigns, space capabilities have further supported other elements of democratisation and good governance. This is salient in policy planning, implementation, and assessment. Prominently, space capabilities have been employed to support election management in Nigeria. Since 2010, Nigeria has made significant efforts toward digitalisation of election planning and processes. In the wake of the 2011 General Elections, the then head of the Independent National Electoral Commission (INEC), Prof Attahiru Jega requested for the assistance of NASRDA in producing an updated map of the country, to guide the commission in revising the geostrategic considerations for the distribution and location of polling units and electoral materials. Satellite data produced by the space agency helped INEC to identified many (new) settlements that were not captured in the initial elections, and reassess the demographic growth across the country. These measures contributed to the efforts to increase political participation, representativeness of government and entrench democracy in the country.

To reduce irregularity in electoral management in Nigeria, INEC also initiated e-platform that allows for instant transmission of all results from different parts of the country to the election situation monitoring room in its HQ, Abuja.

This e-platform like many other public platforms is supported by both terrestrial and space assets for telecommunication. Accordingly, INEC has partnered with NIGCOMSAT Ltd for IT support in election management, especially in satellite support for relevant software, telecommunication and connectivity, since 2011 (Adaramola 2010; Ugbede 2018). This is important to ease the administrative demand of managing the massive number of polling units and personnel engaged by INEC during elections. Notably, the INEC engaged about one million ad-hoc staffs across 119,974 polling units in the 2011 and 2019 General Elections, and 176,846 polling units in the 2023 General Elections (Omotola and Oyewole 2023). In this case, space capabilities supported some of the efforts to entrench transparency in electoral management, and checkmate fraudulent dealings and manipulation that characterised the traditional, manual and offline transmission of results in the many elections in Nigeria.

Space support for public policy planning, implementation, and evaluation have also contributed to improvement in government efficiency. The Nigerian satellites have been used to produce a series of surveys of land use and land cover, which has been supplied to the Federal Capital Authority and many state governments to support urban and regional planning across the country. These data among others have improved town planning and modernisation in some parts of Nigeria, where state and local governments have developed institutional capacities to design and implement relevant strategic plans. Satellite data have equally supported ministries of transportation, agriculture, education, and health among others, as well as their various departments as a contribution to policy efficiency and good governance in Nigeria.

Despite the foregoing prospects of space support for democracy, elections and governance in Nigeria, it is not a guarantee for free and fair elections, sustainable democracy and efficient governance. As evident in 2023 General Elections, where electronic transmission of electoral results was first fully deployed by INEC, technical and possibly human factors undermine the credibility of the whole process. While the results for elections of national assembly members were uploaded instantaneously in many polling units and collation centres, the results of the Presidential election that was conducted at the same time failed to upload in most polling units. This fuelled criticism of the electoral body, especially among the opposition parties and their supporters, and scepticism against the integrity of the results declared in the final analysis.

Many government institutions in Nigeria also underutilised space support to the detriment of their effectiveness to promote human development and security. Notably, department of town planning in many parts of Nigeria lacked the institutional capacity to design and implement community and urban development plan, even with the space capacities and capabilities of the country (Oyewole 2021a). Some of the institutions have little or no contact with the space agency, where they can secure relevant satellite with little or no charges. Similarly, many of the Nigerian government ministries, departments and agencies in agricultural sector lack the capacity to formulate sound policy, as they can't accurately determine or project the size and dynamics of agricultural investment and productivity (Amusan and Oyewole 2023). Yet, their adoption and utilisation of space support remain largely inadequate to guarantee food security.

5.3.3 Food, Water, and Environmental Security

Space capabilities have been employed to support research and policy quests for food, water, and environmental security in Nigeria. The water and the environment are critical factors in food production. Hence, they cannot be ignored in food security calculation of a country. Nigeria has sponsored a series of space supported R&Ds in this area. Satellite data have helped the country to determine the growing desertification in the North, as well as forecast and prepare for period of drought. A series of space supported studies have been conducted to determine Nigeria's basement complex, as well as availability and accessibility of freshwater across the country (see for example Olusola and Fashae 2019). These among other have supported policy planning on agricultural, domestic and industrial utilisation of water, with significant implication for food production and importation, in the country.

It is notable that the percentage of freshwater used out of the available internal resources of the country rose from 2.529% in 1987 to 4.665% in 2002 and 5.642% in 2012. Amidst these, the available data from the World Development Indicators dataset show that the proportion of freshwater withdrawer for agriculture in Nigeria has declined over the years, as population growth continue to increase domestic withdrawer. Freshwater withdrawer for agriculture declined from 70.13% of the total in 1987 to 53.44% in 2000 and 47.62% in 2005, while domestic withdrawer rose from 20.13% in 1987 to 31.45% in 2000 and 35.43% in 2005. Within this period, industrial withdrawals were between 9 and 16%. In 2010, agriculture, domestic and industrial withdrawal of freshwater were 44.19%, 40.1% and 15.76% of the total withdrawal respectively (World Bank 2020). Behind this trend is an alarming decline in irrigation system to support agriculture and food production. These among other things can help us to explain the reason why food importation of the country was alarming in 2011, when food imports were 30.56% of the total national mechanise imports, the highest ever since 1960 (see World Bank 2020). In these among other critical situations, space support for relevant policy cannot be totally ignored.

Nigerian government is faced with a challenging task of determining cultivated land, planted crops and expected yield in a vast country that is characterised with small scale farming and weak institutional capacity of the local government to generate relevant and reliable data that are necessary for policy planning for agricultural and food production, importation and exportation. In this case, satellite data offers a very useful basis for policy planning. Satellite data is also relevant to determine and forecast weather, climatic condition, soil nutrient and fertility, water availability, pest population, fish and other maritime population among other things that enhances food production and ensures precision agriculture (Amusan and Oyewole 2023). In this manner, the space agency has generated and archived numerous satellite data that can be accessed and utilised by government agencies, research institutions and private firms and individuals to advance their interests in the agricultural sector with little or no cost.

Space support for terrestrial communication infrastructures has enhanced the connectivity of the remote rural and coastal communities, which are very critical to farming and fishing as well as food security consideration of the country. This is important to keep these food producers in remote locations up to date about government policy (especially on support for farmers and tariff policy) and the general market values of their products. In this manner, the federal government under President Jonathan adopted an e-wallet platform to register and support fair distribution of subsidised fertilizer to farmers, with the aim of reducing the role of middlemen and attendant room for corruption, in Nigeria in 2014 (Manyika 2016; Oyewole 2017a). The growth of e-commerce has also boosted the market prospects of many farmers in procuring quality inputs (seeds and fertilisers) and machineries, and selling finishing goods beyond their immediate communities. In this case, satellites have supported remote telecommunications.

The utilisation of space support for environmental security generally in Nigeria has recorded its progress and challenges. Satellite data have supported research and policy responses against deforestation and soil erosion in Nigeria, with most emphasis in the South. Research and policy responses against desertification of the North, as well as depletion of Sambisa Forests have benefited from Nigeria's space policy, programme and capabilities. Shrinking of Lake Chad and other water bodies, most especially Kanji and other dams that provide the country with hydro-electric power, have been studied with space support. Satellite data have equally supported Nigeria's search for oil reserve in the Lake Chad as well as River Benue and Niger Basins. The same assets have been employed to assess oil spillage and gas flaring in the Niger Delta region as well as their environmental and health implication. In all these and other cases, space support is relevant for research as well as policy advocacy, consideration, adoption, implementation and assessment of environmental security in Nigeria (see Ajonye et al. 2016; Anifowose 2019; Ayodele et al. 2013; Bahago et al. 2019; Ikpaya et al. 2016; James et al. 2007, 2013; Mahmoud et al. 2016; Olusola and Fashae 2019; Oyewole 2022c).

Despite its promises, space supported policy research and advocacy for environmental security does not always translate to adequate policy attention, adoption and implementation in Nigeria. Although satellite data that are relevant to forecast weather, soil fertility, water and pest among other things that can enhance food production and ensure precision agriculture are increasingly available in Nigeria, they remain largely inaccessible for majority of the farmers, who have largely continued to maintain their small holding and employ primitive farming strategies. Similar data on maritime ecosystem are inadequately used by government, most especially at the state and local levels, to develop coherent policy to prevent or reduce overfishing. At the same time, relevant satellite data remain largely inaccessible to majority of the people that are involved in fishing to enhance productivity and protect endangered maritime species.

5.3.4 Quality of Health System

Space policy and the attendant capabilities have contributed to the quality of health system in Nigeria. At the basic level, many Nigerians across the country, including those in remote areas, can now get some tips for healthy living through public information on food, water, drugs and environment from satellite TV and Radios as well as internet. In this manner, those in remote locations now have the opportunities for timely access to information on food, drink and drugs that have been declared unsafe for public consumption, as well as warning for members of the public on environment disasters, contaminated areas and outbreak of diseases. Prominently, public responses to the threats of Ebola in Nigeria include satellite-based broadcasting, which were critical in the efforts to educate those living in rural and remote communities, especially in warning them against consumption of bush meet, which may be contaminated. Satellite supported platforms are also useful in a series of public engagements for sustainable basic education on healthy and resilient living in many remote communities that are affected by oil spillage and gas flaring in the Niger Delta region as well as lead poison in many illegal mining sites in the North-West.

Telemedicine is another important area with space support for quality health system. As revealed by an insider, the Nigerian civil space institution and assets were first employed to support telemedicine as a pilot project in early 2000s. As at that time, however, the issue of telemedicine was not operationalised beyond pilot stage. Nevertheless, public and private interests and investment in this programme have been revived in recent years. One of the most important private actors in this area is the Lafiya Telehealth App Ltd., which is a subsidiary of BeepTool. However, the outbreak of coronavirus (COVID-19) in Nigeria made the development, deployment and utilisation of telemedicine a necessity, as the government of the country like others across the world struggle to enforce social distancing to contain the spread of the virus (Obikunle 2020; Oludimu 2020). During this period, NASRDA and Lafiya Telehealth App Ltd launched space supported telemedicine programme (NASRDA 2020b; Space in Africa 2020). This and other space supported initiatives contributed to the efforts of the country to contain the spread of COVID-19, supports connectivity among coordinating and operational centres and response teams, and determine, contact, quarantine and support some of those that were infected by the virus.

Satellite data have equally supported research and policy responses with large scale mapping to determine outbreak and concentration of diseases. They were used by many researchers to mapped out the spread of Ebola in West Africa in 2014. Similar data have been useful for mapping spatial distribution and concentration of swamp and other conducive environments for mosquitos in malarial research as well as related policy advocacies and considerations in Nigeria and beyond. Moreover, satellite map and, in most cases, GPS have supported health workers with details to track the location of those in emergency medical situations with appreciable precision. These have been very useful in responses to COVID-19. Nevertheless, institutional capacities, preparedness and strategies can best explain the gap between the potentials and realisations of space support for effective health policy in Nigeria.

5.3.5 Quality Education for All

Nigeria's space policy and capabilities have supported the quest for quality education for all in the country. Terrestrial and space infrastructures have supported e-learning through internet, TV and Radio. Public education has been sponsored by the government and NGOs through these platforms across the country. During the COVID-19 lockdown in Nigeria, some states and private institutions utilised satellite supported platforms for basic education. Many public primary and secondary school programmes ran on dedicated local and satellite TV and Radio stations, while many private schools embraced internet supported e-learning across the country. Prior to these, the high degree programmes of the National Open University of Nigeria (NOUN) and many regular universities, especially those with distant learning programmes, have been based on e-learning platforms. These have brought the opportunities to enrol for higher education to many people, including those in the remote parts of the country. In these and many other ways, education have become more accessible to the population of the country. Alas, very few universities and other tertiary institutions, which are mostly privately owned, are able to operate fully on e-learning in Nigeria during COVID-19, many others were closed down.

The space programme has equally boosted the quality of education in Nigeria. The number of trained personnel with PhD and MSc in the industry have supported the capacity of the country in STEM, and particularly in space S&T education. Besides, the space agency has supported R&D and the improvement in the quality of education in Nigeria with institutional and individual grant to develop their capacity in space S&T. It was revealed that between 12 and 15 universities received this kind of grant between 2004 and 2008. Amidst these is FUTA, which created CESRA with its first grant. The university and its centre equally received another grant from the agency, which was utilised to support the construction of Nigerian EduSat-1 satellite. With these among other initiatives, FUTA is proud to have produced many space scientists and engineers, including the founder and director of Space in Africa. The space agency and its centres also offer internship opportunities to Nigerian students, as well as training and re-training of academics and professionals from across the country and the continent.

Nigerian space assets have equally provided research centres, universities and other high institutions of learning with space-related data for R&D across the country. It is important to note that satellite data has been supplied to universities and research institutes across the country without charges. NASRDA donated more than 4,000 satellite images taken by its first satellite NigeriaSat-1 to Nigerian universities and research institutions. These images were estimated to worth 3 billion Naira (about US$8.3 million). It also donated 4.5 billion Naira (US$12.4 million) worth of images from NigeriaSat-X to 35 universities across the country (Space

Table 5.1 Nigeria's contribution to knowledge on space science and technologies, 1996–2019

Fields	No of publications	Citations	Ranking in Africa	Ranking globally
Aerospace engineering	335	1,718	5/43	56/187
Astronomy and astrophysics	386	2,679	3/38	58/177
Atmospheric Science	618	7,179	2/53	42/217
Control and System Engineering	654	2,668	6/47	71/188
Global and Planetary Change	279	4,126	4/55	47/217
Physics and Astronomy (miscellaneous)	2,076	9,622	5/55	61/206
Space and Planetary Science	496	4,232	3/48	58/205

Source Compiled by the author from Scimago (2020) database

in Africa 2019a). These among others represent space support for education to ensure quality of the system. The end products of these and related developments are evident in Nigeria's growing contribution to the body of knowledge on space S&T as considerably presented in Table 5.1.

As evident in Scimago ranking, Nigeria contributed 355 scholarly publications, which attracted 1,718 citations, to knowledge on aerospace engineering between 1996 and 2019, making the country 56th out of 187 globally and 5th out of 43 in Africa in this consideration. At the same time, Nigeria contributed 385 scholarly publications, which attracted 2,679 citations, to the body of knowledge on Astronomy and Astrophysics, making the country 58th out of 177 globally and 3rd out of 38 in Africa. Nigeria contributed 618 publications, which attracted 7,179 citations, to knowledge on Control and System Engineering, making the country 71th out of 188 globally and 6th out of 47 in Africa. With 279 publications in the field of Global and Planetary Change that attracted 4,126 citations, Nigeria was 4th out of 55 countries in Africa and 47th out of 217 globally. The country came 5th out of 55 countries in Africa and 61st out of 206 globally with 2,076 publications in Physics and Astronomy (miscellaneous). In the field of Space and Planetary Science, the country contributed 496 publications that fetched 4,232 citations to rank 58th of 205 globally and 3rd of 48 in Africa within the period under consideration.

Despite the foregoing claims on the nexus between space and education, there are enduring concerns that the knowledge of space, its attendant technologies and their utilisation are limited in Nigeria. Education still remains out of the reach of many Nigeria, despite the growing space support for internet penetration and the resultant e-learning, mainly due to poverty, inequality, culture and tradition, and government policy failure, inadequacies, inconsistencies and insensitiveness. Equally, the quality of internet and other scientific data access in Nigeria's education institutions does not reflect the space and other relevant investment, capacities and capabilities. The inadequate motivation of scientists and technologists have also limited the impacts of space support for education and research in Nigeria. The attention of social scientists, and political and policy scientists particularly, on the subject matter are even abysmal. These among others show the utilisation and underutilisation of space R&D for education in Nigeria.

5.3.6 Poverty Alleviation and Employment Opportunities

Space support for development in Nigeria has directly and indirectly contribute to efforts aimed at alleviating poverty and reduce unemployment in Nigeria. Thousands of Nigerians are directly employed in the space industry, especially in the public institutions. Satellite supported tele-communication, most especially subscription-based TVs and Radios, have created a chain of employment opportunities in the country. Space support for agriculture and food production have also alleviated poverty in Nigeria. This is a sector that is responsible for 36.38% of the total employment of the country in 2019 (World Bank 2020). Space support for food, water, and environmental security as well as health system is also relevant for heathy leaving and poverty alleviation in Nigeria. This is important in a country, where 3.5% of the population were pushed below international poverty line of US$1.90 per day due to out-of-pocket health care expenditure (World Bank 2020). Besides, food poverty in Nigeria was officially reported to be as high as 75.7 in 2009/2010 (NBS 2017). In these and other ways that were identified, space support have contributed indirectly to some poverty alleviation efforts across Nigeria, even the idea and extend of poverty alleviation in country remain generally contentious.

Despite the potential for space support, poverty level in Nigeria remains alarming. More than 20% of the population are below poverty line of US$1.9 per day (World Bank 2020). Corruption, mismanagement of public resources, insincerity and irresponsiveness of public officials are major challenges against policy efforts to alleviate poverty in Nigeria. Recently, government policy on currency value and designation, and petroleum subsidy have forced many more people below the poverty line. Equally, there is a concern that space supported technological advancement and economic transformation will create a socioeconomic condition that will necessitate smart workforce and less job opportunities, especially for a significant proportion of the population that are unskilled and semi-skilled labours, in Nigeria. These are potentially dangerous for national security that is currently threatened by youth unemployment.

5.3.7 International Cooperation

Beyond the struggle for competitive advantage, Nigeria's space policy has fixed the country in the liberal universalist dream of using space and attendant technologies to advance international cooperation and benefits for humanity. In this framework, the people and government of Nigeria have benefited from the space capabilities of other countries, as well as used their capabilities to benefit others as well. The space policy and programmes of the country have benefited from the support of many countries that have helped in technological transfer and services. Several companies from the USA, the UK, France, the UAE, Netherlands, Sweden, and Luxemburg have secured space station permit in Nigeria (NCC 2020). These companies are offering space support services in Nigeria. This is critical in supporting telecommunication industry, including satellite-based TVs, Radios, and internet, as well as in remote sensing of the Earth for various R&D purposes. However, some local stakeholders considered the involvement and activities of these companies in Nigeria as unfair competition for internal capacity building.

The efforts of Nigeria in space institutional capacity and local technological development have been supported by many countries across the world. In training of personnel and human capital in space R&D, countries like the UK, China, Japan, Russia, the USA, Ukraine, and South Africa are prominent supporters of Nigeria. Previous satellites that were sponsored by Nigeria were developed by different companies or institutions or using facilities in the UK, China and Japan. All these satellites were launched in to orbit with the support of Russia, the USA and China. While Japan has supported Nigeria with grant for the development of research satellite and astronomical facilities, China offered the country loan to acquired communication satellites for commercial purposes. These among other things have shown the beneficiary dimension of international cooperation in space R&D in Nigeria. However, the actual extent of space and related technological transfers and adoption in these processes remain debatable.

Nigeria is involved in data sharing cooperation with members of the Disaster Management Constellations (DMC), which are made up of countries (Algeria, China, Turkey and the UK) with satellites built by SSTL (Curiela et al. 2005; Mohammed 2011). This is similar to the data sharing arrangement from constellation of satellites with Algeria, Kenya, and South Africa under the ARM, which was agreed in 2003 (Mostert and Jacobs 2008; Oyewole 2017a). Nigeria is also one of the leading supporters of space cooperation in the continent, which has produced African Space Policy and its institutionalisation with the establishment of the African Space Agency (ASA). Nigeria is equally a member of the Regional African Satellite Communication Organisation (RASCOM) and the United Nations Office for Outer Space Affairs (UNOOSA). These among other platforms have allowed the Nigerian space programme to benefit many other countries and their peoples.

Imageries generated by the Nigerian satellites have been delivered to many countries to support them in disaster management without charges. Beneficiaries of these data include member of DMC and African Resource Management (ARM) constellation, as well as the US, Australia, and South Eastern countries. As earlier identified, the Nigerian satellites have supported peace-keeping and disaster management in many African countries. The African Regional Centre for Space Science and Technology Education-English (ARCSSTE-E), which is funded by Nigeria, is also instrumental for training many from the English-speaking African countries in space R&D.

5.4 Conclusion

This chapter examines the extent that Nigeria's space policy has been operationalised vis-a-vis the capacities and capabilities of the country, considering the institutions as well as the resources, including human and material (funding and facilities) resources that are committed to the realisation of the set goals. Accordingly, this chapter shows records and gaps in space support for different indicators of (national and human) development and security in Nigeria, with attention for national competitiveness, economic growth and transformation, local technological capacity, defence, security, power projection, disaster management, human rights and freedom, democracy and good governance, food and water security, sustainable environment, quality of health and education systems, and international cooperation.

It is observed that space policy and programme have considerably advanced Nigeria's national and human development and security. However, Nigeria's space capacities and capabilities have been underutilised in many considerations to the detriments of their potentials to contribute the advancement of the development and security interests of the people and government.

References

Adaramola, Z. 2010. 2011: INEC to partner with NIGCOMSAT On ICT. *Daily Trust*.

Adebajo, A. 2017. *The eagle and the springbok: Essay on Nigeria and South Africa*. Johannesburg: Fenele.

Adeshina, R.A. 2002. *The reversed victory: The story of Nigerian military intervention in Sierra Leone*. Ibadan, Nigeria: Heinemann Educational Books.

Aina, F., J.S. Ojo, and S. Oyewole. 2023. Shock and awe: Military response to armed banditry and the prospects of internal security operations in Northwest Nigeria. *African Security Review*. https://doi.org/10.1080/10246029.2023.2246432.

Ajonye, S.E., et al. 2016. Remote sensing assessment of Jabi Lake and its environs: A developmental perspective. *Archives of Agriculture and Environmental Science* 1 (1): 1–8.

Akinrinde, O.O., and S. Oyewole. 2021. Neo-colonialism and developmental challenges of post-colonial Africa. *Jurnal Pertahanan* 7 (3): 398–410.

Akinyede, J.O., and R.A. Boroffice. 2013. *Nigeria's Quest in Space*. Ile-Ife: Obafemi Awolowo University Press.

Akinyemi, M. 2020. Satellite is contributing only 0.2% to Nigeria's internet connectivity—NCC. *Space in Africa*.

Al-Ekabi, C., and S. Ferretti, eds. 2018. *Yearbook on space policy 2016: Space for sustainable development*. Geneva: Springer International Publishing AG.

Al-Ekabi, C., B. Baranes, P. Hulsroj, and A. Lahcen, eds. 2017. *Yearbook on space policy 2015: Access to space and the evolution of space activities.* Geneva: Springer.

Amnesty International. 2018a. Nigeria: Analysis of the air force raid in Adamawa state.

Amnesty International. 2018b. Nigeria: Dozens killed as military launches air attacks on villages beset by spiralling communal violence.

Amusan, L., and S. Oyewole. 2017. The quest for hegemony and the future of African solution to African development problems: Lessons from headways in African security sector. *Journal of Asian and African Studies* 52 (1): 21–33.

Amusan, L., and S. Oyewole. 2023. Precision agriculture and the prospects of space strategy for food security in Africa. *African Journal of Science, Technology, Innovation and Development* 15 (3): 325–336.

Anifowose, A. Y. B. 2019. Remote sensing analysis of the evolution and retreat dynamics of the Auchi Gully, Southwestern Nigeria. In *Earth observations and geospatial science in service of sustainable development goals: 12th international conference of the African association of remote sensing and the environment*, ed. S. Wade. Cham: Springer.

Ayodele, F.M., et al. 2013. Residential-source emission inventory for the Niger Delta—a methodological approach. *Journal of Sustainable Development* 6 (6): 98–120.

Bahago, R. A. et al. 2019. Analysis of the 2012 flooding events downstream of Shiroro reservoir, a case of Gurmana Niger State, Nigeria. In *Earth observations and geospatial science in service of sustainable development goals: 12th international conference of the African association of remote sensing and the environment*, ed. S. Wade. Cham: Springer.

Burger, E., and G. Bordacchini, eds. 2019. *Yearbook on space policy 2017: Security in outer space: Rising stakes for civilian space programmes.* Geneva: Springer.

Carnimeo, N., and S. Oyewole. 2020. Noi Contro I Petropirati Del Golfo di Guinea. *L'italia È Il Mare* 11: 297–304.

Curiela, A.S., L. Bolanda, J. Cooksleya, M. Bekhtib, P. Stephensa, W. Suna, and M. Sweeting. 2005. First results from the disaster monitoring constellation (DMC). *Acta Astronautica* 56 (2005): 261–271.

Daxecker, U., and B. Prins. 2021. *Pirate lands: Governance and maritime piracy.* Oxford: Oxford University Press.

Froehlich, A., and A. Siebrits. 2019. *Space supporting Africa: A primary needs approach and Africa's emerging space middle powers.* Cham: Springer.

Harding, R.C. 2013. *Space policy in developing countries: The search for security and development on the final frontier.* New York: Routledge.

Harvey, B., H.H.F. Smid, and T. Pirard. 2010. *Emerging space powers: The new space programs of Asia, the Middle East and South America.* New York: Springer.

IISS (Institute of International Security Studies). *The military balance, 2020.* London: IISS and Francis & Taylor.

Ikpaya, I. O., S. O. Onuh, C. U. Achem and F. Y. Madalla. 2016. *Quest of Nigeria into space for sustainable development.* American Institute of Aeronautics and Astronautics, Inc.

James, G.K., et al. 2007. Satellite-based assessment of the extent and changes in the mangrove of the Niger Delta. *Marine Geodesy* 30: 249–267.

James, G., H. Shaba, O. Zubair, A. Teslim, G. Yusuf, and A. Nufu. 2013. *Space-based disaster management in Nigeria: The role of the international charter "Space and Major Disasters".* FIG Working Week 2013: Environment for Sustainability, Abuja, Nigeria, 6–10 May.

Lal, B., E. J. Sylak-Glassman, M. C. Mineiro, N. Gupta, L. M. Pratt, and A. R. Azari. 2015. *Global trends in space volume 1: Background and overall findings.* IDA Paper P-5242, Vol. 1 (June). Washington, DC: IDA Science & Te.chnology Policy Institute.

Lopez, L.D. 2015. Space sustainability approaches of emerging space nations: Brazil, Colombia, and Mexico. *Space Policy* 30: 1–6. https://doi.org/10.1016/j.spacepol.2015.12.004.

Luncedo, N., and K. Gottschalk. 2013. The growth of space science in African countries for Earth observation in the 21st century. *South African Journal of Science* 109 (1–2): 1–5.

Mahmoud, M. I. et al. 2016. Analysis of settlement expansion and urban growth modelling using geoinformation for assessing potential impacts of urbanization on climate in Abuja City, Nigeria. *Remote Sensing*, 8(220): 1–24. https://doi.org/10.3390/rs8030220.

Manyika, J. 2016. Lions go digital: The internet's transformative potential in Africa. http://www.mckinsey.com/industries/high-tech/our-insights/lions-go-digital-the-internetstransformative-potential-in-africa. Accessed 19 Jan 2016.

Mistry, D., and B. Gopalaswamy. 2012. Ballistic missiles and space launcher vehicles in regional powers. *Astropolitics* 10 (2): 126–151.

Mohammed, S.O. 2011. *Nigerian space research and development agency. Director-general presentation.* Abuja: NASRDA.

Mostert, S., and M. Jacobs. 2008. ARM constellation—Establishing a regional remote sensing asset. *Acta Astronautica* (63): 221–227.

NASRDA. 2020. NASRDA rolls out telemedicine project to combat COVID-19 in Nigeria.

NBS. 2017. Annual Abstract of Statistics 2016. Abuja: Nigerian National Bureau of Statistics (NBS).

NCC. 2020. List of authorised space station in Nigeria.

Obikunle, F. 2020. Successes, challenges of telemedicine adoption in Nigeria clinical settings. *The Guardian.*

OECD. 2014. *Space economy at a glance 2014.* OECD Publishing.

Ogunnubi, O., and S. Oyewole, eds. 2020. *Power politics in Africa: Nigeria and South Africa in comparative perspective.* Newcastle: Cambridge Scholar Publishing.

Ogunnubi, O., and N. Okeke-Uzodike. 2016. Can Nigeria be Africa's hegemon? *African Security Review* 25 (2): 110–128.

Oludimu, T. 2020. How the COVID-19 pandemic amplifies the need for telemedicine adoption in Nigeria. *Techpoint Africa.*

Olusola, A., and O. Fashae. 2019. River long profiles of selected third-order basins in basement complexes. In *Earth observations and geospatial science in service of sustainable development goals: 12th international conference of the African association of remote sensing and the environment*, ed. S. Wade. Cham: Springer.

Omeni, A. 2018. *Counter-insurgency in Nigeria: The military and operations against Boko Haram, 2011–2017.* London: Routledge.

Omotola, J. S., and S. Oyewole. 2023. *Electoral violence in Nigeria: Insights from off cycle subnational elections.* Asian and African Studies.

Omotola, J.S., and S. Oyewole, eds. 2024. *The political economy of kidnapping and insecurity in Nigeria.* Charm: Springer.

Onuoha, F. C., and S. Oyewole. 2018. *Anatomy of Boko Haram: The rise and decline of a violent group in Nigeria.* Doha: Al Jazeera Centre for Studies.

Oriola, T. B., F. C. Onuoha and S. Oyewole. 2021. Introduction: Boko Haram in the Lake Chad Basin—dimensions, activities and trajectories. In *Boko Haram's campaign of terror in Nigeria: Contexts, dimensions and emerging trajectories*, ed. T. B. Oriola, F. C. Onuoha and S. Oyewole. London: Routledge.

Oshita, O.O., I.M. Alumona, and F.C. Onuoha, eds. 2019. *Internal security management in Nigeria: Perspectives, challenges and lessons.* Singapore: Palgrave Macmillan.

Oyewole, S. 2013. Boko Haram and the challenges of Nigeria's war on terror. *Defense and Security Analysis* 29 (3): 253–262.

Oyewole, S. 2015b. Boko Haram: Insurgency and the war against terrorism in the Lake Chad region. *Strategic Analysis* 39 (4): 428–432.

Oyewole, S. 2016a. Suppressing maritime piracy in the Gulf of Guinea: The prospects and challenges of the regional players. *Australian Journal of Maritime and Ocean Affairs* 8 (2): 132–146.

Oyewole, S. 2016b. The fate of hostages: Nigeria's conflict theatres in comparative perspective. *African Security Review* 25 (2): 193–207.

Oyewole, S. 2016c. Rescuing Boko Haram's schoolgirl victims. *New Zealand International Review* 41 (1): 25–28.

Oyewole, S. 2017a. Space research and development in Africa. *Astropolitics* 15 (2): 185–208.

Oyewole, S. 2017b. Making the sky relevant to battle strategy: Counterinsurgency and the prospects of air power in Nigeria. *Studies in Conflict & Terrorism* 40 (3): 211–231.

Oyewole, S. 2018b. Flying and bombing: The contributions of air power to security and crisis management in the Niger Delta Region of Nigeria. *Defence Studies* 18 (4): 514–537.

Oyewole, S. 2019. The fracturing of pro-Biafra nationalist movements. *African Conflict and Peacebuilding Review* 9 (1): 1–23.

Oyewole, S. 2020a. The quest for space capabilities and military security in Africa. *South African Journal of International Affairs* 27 (2): 147–172.

Oyewole, S. 2020c. Civil-military relations: Conflict and cooperation between military bases and host communities in Nigeria. *African Security* 13 (4): 353–379.

Oyewole, S. 2021b. Struck and killed in Nigerian air force's campaigns: Assessment of air strike locations, targets and impacts in internal security operations. *African Security Review* 30 (1): 24–47.

Oyewole, S. 2023a. The African orbital journey in space age: From spectator to participant. *Vanguardia Dossier* 88 (15): 72–75.

Oyewole, S. 2023c. Piracy and the challenges of Nigeria's maritime domain. *Democracy & Development: Journal of West African Affairs* 7 (1): 55–65.

Oyewole, S., F. Aina, and S.J. Ojo. 2022. Wings over flies: Air campaigns against armed banditry in North West Nigeria. *RUSI Journal* 167 (4/5): 92–103.

Oyewole, S., and A. Olajubu. 2020. Security agencies, lockdown enforcement, and human rights. In *Political economy of pandemics and implication of COVID-19 for Nigeria*, ed. A.O. Sulaiman, 447–463. Abuja: National Institute for Legislative and Democratic Studies.

Oyewole, S. 2015a. Pirate sanctuary and the political economy of piracy in the Gulf of Guinea. M.Sc. dissertation, University of Ilorin, Nigeria.

Oyewole, S. 2018a. Globalisation of Militant Islamist Movements and Insecurity in Africa. Presentation at the 15th General Assembly of the Council for the Development of Social Science Research in Africa (CODESRIA) on *Africa and the Crisis of Globalisation*, between 17 and 21 December. Held at King Fahd Palace Hotel, Pointe des Almadies, Dakar, Senegal.

Oyewole, S. 2020b. Military capabilities of regional powers in Africa: Nigeria and South Africa in comparative perspective. In *Power politics in Africa: Nigeria and South Africa in comparative perspective*, ed. Ogunnubi, Olusola and Samuel Oyewole. Newcastle: Cambridge Scholar Publishing Ltd.

Oyewole, S. 2021a. One of Nigeria's satellites is on its last legs: why this is worrying. *The Conversation.* https://theconversation.com/one-of-nigerias-satellites-is-onits-last-legs-why-this-is-worrying-165068.

Oyewole, S. 2022a. Air and space strategy in African counterinsurgency operations: Insights from the Nigeria's war against Boko Haram. Presentation at the International Conference on Counter-terrorism, Technology and Development in Africa, jointly organised by the Centre for Military Studies, Stellenbosch University, South Africa and the Africa Research Institute, Óbuda University, Hungary, September 22.

Oyewole, S. 2022b. Air and space strategies in African disaster management. Virtual Lecture of the (British) Royal Air Force (RAF) Museum on August 11.

Oyewole, S. 2022c. How space research can help Africa meet the SDGs. Presentation at a webinar of the University of Johannesburg's Institute for Pan-African Thought and Conversation (IPATC) on "Reaching for the final frontier: The African Space Sector".

Oyewole, S. 2023b. Military space strategies and African realism: Egypt, South Africa and Nigeria. In *Oxford handbook of space security*, ed. S. M. Pekkanen and P. J. Blount. Oxford: Oxford University Press.

Oyewole, S. 2023d. Niger's coup weakens regional fight against Boko Haram: four reasons why. *The Conversation.* https://theconversation.com/nigers-coup-weakens-regional-fight-against-boko-haram-four-reasons-why-212202.

Scimago ranking. 2020.

Space Foundation. 2015. *The space report 2015: The authoritative guide to space activities.* Colorado Spring: Space Foundation.
Space in Africa. 2019a. *NewSpace Africa industry report*. Space in Africa.
Space in Africa. 2019b. How much has NASRDA contributed to Nigeria's economic growth and development?.
Space in Africa. 2020. Lafiya telehealth is changing the way medicine works in Africa.
Ugbede, L. 2018. 2019: INEC to partner NIGCOMSAT for successful polls. *Premium Times*.
World Bank. 2020. World development indicators. Available at http://datatopics.worldbank.org/world-development-indicators/. Accessed 10 March 2021.

Part III

Space Policy, Development, and Security in South Africa

Space Politics, Policy and Strategy in South Africa

6

Abstract

The South Africa's space policy and strategy, and the attendant capacities and capabilities have emerged from centuries of the nation's scientific, technological, political and socioeconomic developments, and defence, security and international engagements. This chapter examines the national and international sociocultural, economic, political, historical and military contexts, determinants and interests in space policy and strategy, the attendant capacities and capabilities of South Africa. It examines the trajectories of South Africa's space policy and strategy from the indigenous cultural astronomy and settlers' modernisation of space R&D in precolonial and colonial eras to the amateur rocketry, militarised apartheid nuclear, missile and space triad programme, and the post-apartheid new dawn. It is against this background that this chapter examines extant laws, policies and strategies of South Africa to determine its national interests and perspectives on space and related technologies, socioeconomic development, and security.

6.1 Introduction

This chapter examines space policy and strategic interests and commitments of South Africa in the context of social, economic, political, and historical determinants. Beyond this introduction, this chapter is divided into four sections. The first section provides the background information on South Africa, which is necessary to understand the political, socio-economic, historical and security dynamics of its space policy and strategy. The second section examines the evolution of space R&D in South Africa, paying attention to astronomy and basic space R&D as well as nuclear, missile and space triad programmes, and the historical, political, economic, and social contexts, determinants and dynamics. This is important to understand the third section on the policies and strategies of the country on

S. Oyewole, *Utilitarianism in Outer Space*, Southern Space Studies,
https://doi.org/10.1007/978-3-031-49646-2_6

space, science and technology, socio-economic development, and security. The fourth section provides concluding reflection on space politics, policy and strategy in South Africa. Accordingly, this chapter serves as a background to South Africa's space policy and strategy, which is critical to understand subsequent chapters on space capacities and capabilities, and their contribution to socioeconomic development and security.

6.2 Background Information on South Africa

This section provides a brief background information on the people and government of the Republic of South Africa (RSA). It covers geographical, historical, sociocultural, economic, political, security development of the country. RSA is situated at the Southern tip of Africa, between 22° to 35° South Latitude and 17° and 33° East Longitude. It occupies a surface area, which covers 1,219,602 km^2 (GCIS 2019; CIA 2020). In addition to a coastline of 2,798 km (along the Atlantic and Pacific Oceans), RSA shares 5,244 km of land boundaries with Botswana, Lesotho, Mozambique, Namibia, and Swaziland (CIA 2020; GCIS 2019). The country hosts a number of internal waters and other physical features like beaches, bushveld, coaster plain, desert, forest, grassland, hill, plateau, and mountains (CIA 2020; GCIS 2019; Grab and Knight 2015).

The entity that is now known as RSA today is a product of thousands of years of occupation of the affected territory by black Africans and their ancestors, before the arrival of the whites from Europe and Indians from Asia. The waves of immigration and settlement in the southernmost part of Africa paved the way for the scrabbling, expropriation, and colonisation of South Africa by different European powers (capitalists and states) between seventeenth century and the early twentieth century (Feinstein 2005; Landau 2010; Thompson 2000; Stephens 2003). On May 31, 1910, four British colonies, including Cape Colony, Natal, Transvaal, and Orange Free State, formed the Union of South Africa, which became a republic on May 31, 1961. However, the apartheid regime, a system of socioeconomic and political segregation based on race, was maintained in the country (Beinart and Dubow 1995; Seekings and Nattrass 2005). This was the situation where the black Africans, who constitute majority of the population, were perpetually subjugated to the whites, who constitute the minority, until majority rule was ushered on April 27, 1994, and multi/non-racial constitution was adopted on December 4, 1996 (effective from February 4, 1997) (Clark and Worger 2011; Guelke 2005; Lodge 1989).

The population of RSA was estimated to be 56,463,617 as at the middle of 2020, making it the 26th most populous country in the world (CIA 2020). Official report, however, shows that the country was populated by 58.78 million in mid-2019 (GCIS 2019). At the same time, AfDB et al. (2019) reported 58 million, which constitutes 4.4% of the total population of the continent and made RSA the 6th most populous country in Africa. Amidst these, black African constituted about 80.9% of the population, while the Coloured was 8.8%, the White was 7.8%

and the Indian/Asian 2.5% (CIA 2020; GCIS 2019). The population was structured into 27.94% 0–14 years, 16.8% were 15–24 years, 42.37% were 25–54 years, 6.8% were 55–64 years and 6.09% were 65 years and above in 2020 (CIA 2020; World Bank 2020). The population of the country is considerably literate, given 87% literacy rate of adult of age 15 and above as at 2017 (World Bank 2020). Gross primary education enrolment in RSA was 100.8%, while secondary education was 104.7%, and 22.3% for tertiary education. Educational attainment of age 25 and above people include 0.31% doctorate or equivalent and 1.46% with at least master or equivalent (World Bank 2020). Amidst these, the country's labour force was 22.19 million, which was 29th largest in the world, with 27.5% unemployment rate, in 2017 (CIA 2020; GCIS 2019; World Bank 2020).

With a GDP of US$368 billion, which represents 15.8% of the continental total, in 2018, South Africa is the second largest economy in Africa, a position it has retained since it was surpassed by Nigeria in 2013 (AfDB et al. 2019; World Bank 2020). The economy depends largely on services, which contributed 68% to the GDP in 2017. Equally, industries contributed 29.3% to the GDP, including 13.4% from manufacturing, while the agriculture only contributed 2.6% (AfDB et al. 2019). The country is rich in diamonds, gold, iron ore, platinum, copper, uranium, and silver among others. However, the total natural resource rents only account for about 5% of GDP of the country. Amidst these, general government consumption expenditure was 21.28% of the GDP in 2018. In the same year, the external debt of RSA was 46.3% of the GDP, making it the 19th most alarming case in the continent. The GDP per capital of RSA was US$6,414, which was the fifth highest in Africa as of 2018 (AfDB et al. 2019; World Bank 2020).

RSA has practiced popular democracy since 1994. The republic is structured as a quasi-federal system with a strong central government and nine subnational governments at provincial level (Fessha 2010). The country operates a presidential system of government, and a National Assembly that is made of between 350 and 400 members, which are elected for five years through proportional representation. In addition to these is the National Council of Province (NCOP), which is made of 90 members (10 from each province) (GCIS 2019). The political space of the country has been dominated by the African National Congress (ANC) since 1994, although the opposition parties are growing in significance in recent times. However, political stability and integration as well as development and security of RSA have been threatened to various degrees by corruption, poverty, inequality, youth unemployment, HIV/Aids, criminal and xenophobic violence (Amusan and Mchunu 2017; Amusan 2016; Bond 2000; Hanoman 2018; Oyewole 2023c; Pillay 2014; Schulz-Herzenberg 2007; Seekings and Nattrass 2005; Tafira 2018). These are relevant to understand the development of space policy and strategy of South Africa.

6.3 The Development of Space Policy in South Africa: The Historical, Political and Socio-Economic Perspectives

This section provides some insights into the development of space R&D in the framework of politics and policy in South Africa. It pays attention to the historical, political, and socio-economic contexts, determinants and factors that characterised the development of space R&D of the country. Compared to Nigeria, South Africa has richer, complex, and contested history of active space programmes. In addition to its overt records of civil space programmes, South Africa had record of covert space programme, which was integrated with nuclear and missile programmes, and coordinated for defence and security purposes. These are connected to the complex and contested history of the country's socio-political development. To better understand these dynamics, this section will be subdivided into two: the first one will focus on the genesis of country's civil aspects of space programme, with special attention for astronomy and basic space R&D. The second subsection focuses on the development of space R&D amidst missile and nuclear programmes of the country. Although most details of the political and policy developments on space R&D after the apartheid are reserved for the next sections, the analysis in the section generally covers the period before, during and after the apartheid regime.

6.3.1 Astronomy and Basic Space R&D

Space R&D in South Africa, like in other African countries, have their distanced origin in astrological and cosmological visions of different pre-colonial African societies that occupied the territory, as evident in cultures and traditions of many centuries before the arrival of the Europeans. In a similar Afrocentric narrative to the earlier identified case of Nigeria, the knowledge of space in South African distant past was largely characterised by systems of believe, which included myths and rituals, that revolved around celestial bodies as well as their interactions with Earth and influence on terrestrial affairs, including the origin, development and fate of humanity (Holbrook 2015; Holbrook et al. 2008; Mathenjwa 1999; Urama 2008; Urama and Holbrook 2011). The knowledge produced in this process were largely passed from one generation to another through oral tradition, drawing, painting, music, rituals and myths, until the advent of more systematic writing, which started the documentation of some of these ideas (Smedegar 2007; Urama and Holbrook 2011). The documentation process was further aided with the introduction and spread of Western Education in South Africa.

Modern astronomy started in South Africa much earlier than in other parts of the continent. The strategic position of the country on the map as well as the favourable weather and the abundant natural resources in the region, attracted the attention and influx of European visitors, settlers, investors, and investigators, including in the area of astronomy and related scientific studies. As far back as 1700s, several European scientists had visited the Cape of Good Hope and other parts of modern-day South Africa to conduct research, including in the area of geodesy and astronomy (Evans 1967; Smedegar 2007). Amidst these, Nicolas-Louis de La Caille, a leading French mathematician sponsored by the Royal Academy of Sciences in Paris (RASP), made the first deliberate expedition for astronomical purposes between 1751 and 1753. In this process, he determined the position of 9,800 southern stars and catalogued them, carried out a series of studies that significantly impacted the development of space sciences, and calculated the latitude and longitude of the Cape (Evans 1967; Dubow 2019).

With the defeat and occupation of Holland during the Napoleonic War, political control of Cape passed from the Dutch to the British, a process that was completed with a treaty of 1814 (Evans 1967). With the rising British among other European interests in scientific edge amidst accruable political and economic benefits from the region, Royal Observatory was establishment at the Cape by the government of the United Kingdom in 1820 (Cillie 1973; Dubow 2019; Evans 1960, 1967). In 1834, Sir John Herschel—who was earlier credited with the discovery of planet Uranus—with his wife and assistant, Margaret, arrived at the Cape, and soon established a private observatory. At the same time, he was elected as the pioneer President of South African Literary and Scientific Institution (LSI) in 1834. With a 20-foot reflector telescope, Herschel identified over 1,700 nebulae and 2,100 double stars, which had notable implications for the development of theoretical understandings of our Galaxy, the Milky Way (Dubow 2019). Together with Thomas Maclear, the second head of Royal Observatory at Cape, Herschel equally studied comets (Dubow 2019; Lastovica 1995). These developments laid the foundation of modern space R&D in South Africa.

As of 1910, when the Union of South Africa was formed, the country had already become a notable hotspot in the global map of astronomy and related R&D. The geographical location of the country in Southern Hemisphere, which is primarily comparable to that of Australia, New Zealand, Argentina, and Chile, is advantageous in astronomy, given the concentration of relevant facilities in Northern Hemisphere. The dominion status of the country until 1961 also made it an important asset to the British empire. These among others attracted the influx of British among other European and later American space scientists and investors to the country. In this process, local capacity, mainly of the White settlers, was developed in space R&D. Prior to 1961, when the country attained Republican status under White domination, several astronomical observatories were already established. Beside the Royal Observatory at the Cape, there were Leiden Southern Observatory; Boyden Observatory at Bloemfontein, which was controlled by a consortium of five European nations; as well as a telescope of 74-inches reflector of the Radcliff Observatory at Pretoria (Cillie 1973; Evans 1960). Few American based institutions like Harvard and Yale also collaborated, and supported some projects, with some of these observatories, as far back as 1920s and 1930s (Cillie 1973).

As at 1950s, South Africa was at the heart of the conception of Southern Skies in northern hemisphere and associated international collaboration in astronomy. The launch of sputnik by the Soviet Union, and the associated Cold War implications for the USA, also drew NASA close to South Africa in late 1950s and 1960s.

Accordingly, a 26-m parabolic antenna, which was capable of deep space tracking of satellites and space missions, situated at Hartebeesthoek (HartRAO) was operated by South Africa for NASA from 1961. It equally collaborated with the Centre National D'études Spatiales (CNES), the French space agency, to operate a satellite tracking station at Paardefontein from 1965 (Dubow 2019). These strategic partnerships boosted the prospects of radio astronomy in South Africa. In this way, Very Long Baseline Interferometry (VLBI) was installed in HartRAO, with the support of the USA and France. Between 1960s and early 1970s, the British governments and scientific community also played important role to expand South Africa astronomical facilities as well as utilised and maintained the existing ones. These among other collaborations accounted for rapid development in basic space R&D in South Africa (Dubow 2019; Warner 2008).

Despite the foregoing progress, the global wave of nationalism and associated international campaigns against the apartheid regime of South Africa undermined the developmental trajectory of astronomy and basic space R&D in the country. In this way, many international partners were forced to withdraw their support for space R&D in South Africa. This was the case of partner universities in Sweden in 1966 as well as those in Western Germany and many in the USA and other European countries in 1971 (Dubow 2019). NASA withdrew its presence in the country thereafter. These among other things created several vacuums in South Africa, to the benefits of its rival in Southern Skies, that is, Australia and to a lesser extent Chile. In response, the apartheid regime raised local capacity to operate existing facilities, while it maintained, and entered into new secrete agreements for scientific cooperation with few willing countries, of which France and Britain are prominent among them (Dubow 2019).

As detailed in the subsequent section, apartheid South Africa's space policy and programme became subsumed into, dominated by, and reduced to the pursuit of, defence and security interests, which was largely operationalised with the quest for space, missile, and nuclear weapon capabilities, between 1960 and 1980s. Nevertheless, the transition and post-apartheid governments of South Africa considerably demilitarised space programme in the country, and reinstated its R&D motives (Gottschalk 2010; Martinez 2009, 2016; Munsami 2014; Van Wyk 2009). These among others have repositioned astronomy in national development, and South Africa for competitive advantage in Southern skies and associated international collaboration and investment. These are evident in many ambitious projects, of which the South African Large Telescope (SALT) and Square Kilometre Array (SKA) are prominent examples in this context (Dubow 2019; Gastrow 2015; Gastrow et al. 2016; Gastrow and Oppelt 2019; Martinez 2009, 2012a; Twidle 2019; Van Wyk 2009; Walker and Chinigò 2018).

6.3.2 Space, Missile and Nuclear Programmes

Beside astronomical observatories, South Africa also features in other aspects of space R&D, that is, rocketry for space launch and missile capabilities and satellite systems, earlier than many other African countries. The space age, which started with the successful orbiting of the first artificial satellite of the Earth by the USSR in 1957, was made possible with the breakthrough in rocket science and engineering, from where missile technology and orbital vehicles were developed. The age of modern rocket science and engineering emerged between 1990 and 1940s, amidst which 1930s marked the beginning of major rocket tests and flights in many of the leading countries in this consideration (Chun 2006; Van Riper 2004). During the Cold War, this became a quest for trio-capabilities, that is, satellite, missile/space launcher and nuclear weapons, for many countries. South Africa was not left out of this development. Between 1960s and late 1980s, the Apartheid regime developed interests, institutions and policy programmes for missile/space launcher, nuclear weapon, and reconnaissance satellites, with varying, but remarkable, degrees of success (Albright and Stricker 2016; Gottschalk 2010; Howlett and Simpson 1993; Purkitt and Burgess 2005; Van Wyk 2014a; UNDDA 1991).

Rocket, space flight and missile programmes of South Africa can be traced to private civilian research and test flight that dated back to late 1940s. Desmond Prout-Jones, the founder of the South African Rocket Research Group (SARRG), established in 1957, started to test-flight rocket in the country in 1947. South African Interplanetary Society (SAIS) that was established in 1953 was another major player during this period. Between 1959 and 1963, the SARRG launched 528 rockets, which include two-stage, five-stage, supersonic two-stage and liquid-propelled rockets, and 102 firing tests, with varied degree of success. The final rocket launched by the group reached a 40 km altitude, and a 100 km altitude flight was planned before government explosive inspectors banned private rocketry (Gottschalk 2010; Oyewole 2017). Following the ban, the apartheid regime commenced a missile R&D programme in 1963, from where the aborted plan to develop space launch vehicle was conceived in 1980s. During this period, the South Africa space programme was subsumed under defence policy and designed in close connection to the missile and nuclear weapon programmes of the country (Albright and Stricker 2016; Van Wyk 2014a; UNDDA 1991).

As the wave of independence became obvious across Africa in 1960, the apartheid regime in South Africa was confronted with a series of internal and external security challenges. Internally, the regime was confronted with Black resistance to the imposed system of racial domination of the indigenous peoples that are majority under the White minority rule (Clark and Worger 2011; Guelke 2005; Lodge 1989; Mandela 1995). The collapse of colonial powers in southern Africa, especially Portuguese empire, replaced friendly government of neighbouring territories with hostile nationalist governments in the region. Accordingly, the internal resistance, which has both civil and militant dimensions, draw massive support from multiple post-colonial states in Africa and the Third World, who were driven by nationalist governments, as well as the Soviet Union and other communist countries. These among others challenged the security, defence, and regional power projection of apartheid regime in South Africa (Jaster 1989; Stapleton 2010; Thomson 2008; Warwick 2009). In response to internal resistance, the regime adopted militarised policing against the Black population and counter-insurgency against their political movements, with limited success. These among others informed the barn on explosive and civilian rocket programme in 1960s. However, the apartheid space, missile and nuclear programmes were aimed to boost the defence posture of the country and serve as strategic deterrence against frontline states and their ambitious allies, especially the Soviet Union and Cuba (Albright and Stricker 2016; Goodson 2012; van Wyk 2010).

The foundation of South Africa's nuclear research and weapon programme was laid in 1940s and 1950s, when the country was contacted for uranium exploration and supply by the UK and later the USA, who were then interested in the global mapping and control of the deposits, exploitation, and market of this precious product (Albright and Stricker 2016; Purkitt and Burgess 2005; Masiza 1993). Accordingly, South African Atomic Energy Board (AEB) was established in 1948 and national nuclear R&D plan was adopted in 1959. With varying degrees of support of the USA, Britain, France, Germany, Netherlands, Norway, and Israel among other, South Africa developed necessary human resources and infrastructure for its nuclear reactors and further research in this area in the 1960s (Albright and Stricker 2016; Liberman 2004; Purkitt and Burgess 2005). These largely involved technological transfer un/knowingly that it will be used for military purposes. As the regional security dynamics became more and more hostile to the apartheid regime in the 1960s and 1970s, South Africa initiated a process to militarised and weaponised its nuclear R&D (Albright and Stricker 2016; Burgess 2006; Cochran 1994; Goodson 2012; Purkitt and Burgess 2005; van Wyk 2010). Amidst these, Pieter W. Botha played a critical role, first as Defence Minister (1966–1978) and later as Prime Minister (1978–1984) and President (1984–1989) of the apartheid government of South Africa.

In 1977, the first South African nuclear gun device was produced, and the Armaments Corporation (Armscor), which has been the statutory department for arm-procurement of the South African Defence Force (SADF), prepared for its test. However, the launch site and the preparation around the facility were uncovered by the Soviet satellites, couple with details from its intelligence within SADF (Bidgood 2016; Polakow-Suransky 2010; Masiza 1993). The USSR shared the details of this intelligence with the government of the US, which also carried out independent efforts to verify the report, which all played critical roles in the international pressure that was mounted against South Africa for denuclearised (Asuelime and Adekoye 2016; Bidgood 2016; Moore 1987; Polakow-Suransky 2010; Van Wyk 2007). Despite the international pressure, the apartheid regime made further progress in its R&D of nuclear weapons between 1977 and 1990 (Albright and Stricker 2016; Burgess 2006; Masiza 1993; Thomson 2008; UNDDA 1991). In 1993, South African President F.W. de Klerk acknowledged that the country spent about 800 million rand (about US$400 million) on its nuclear weapon programme, and built six nuclear devices and almost completed the seventh device (Masiza 1993; Purkitt and Burgess 2005).

Recognising the challenges of air power in delivering nuclear strike generally, and the limited air power capabilities of the country as well as the then rising air defence capacities of Angola and some other frontline states, the regime also prioritised militarisation of rocket R&D for long-range missile and nuclear delivery system (Anderson and Bell 2019; Baxter 2012; Bin, 1997; IISS 1970–1989; Stapleton 2010; Van Wyk 2014a). With the initial support from France and enduring close collaboration with Israel, South Africa recorded significant progress in its missile programme. The country started production of short-range tactical missiles in mid-1960s, from where Valkiri surface-to-surface artillery rocket with 22 km-range and V3 Kukri air-to-air missile that ranged from 4 to 10 km, were inducted to SADF service in the 1980s (UNDDA 1991). Between 1970 and 1980s, the apartheid regime in South Africa acquired prototypes and production details of Jericho-2, a missile of 1,300 km range and 1,000–1,500 kg payload capacity from Israel, which was converted to space launch vehicle by the Jewish State in late 1980s (Harris et al. 2004; Liberman 2004; Gottschalk 2010; Polakow-Suransky 2010). Amidst these, South Africa developed a series of ballistic missiles –including intermediate range RSA-1 and RSA-2, with warhead mass of 1,500 kg and trajectories of 1,100 km and 1,900 km respectively—and RSA-3 and 4, which were planned and designed on Israeli model to serve as space launch vehicles, were at different stages of development (Gottschalk 2010; Van Wyk 2014a; UNDDA 1991).

In March 1966, the South African Council for Scientific and Industrial Research (CSIR) was commissioned by the Ministry of Economic Affairs and Technology to carry out a feasibility study of totally indigenous space programme. When the details of the report were made public in 1989, it was observed that CSIR recognised that the country has the required human and technical capacities, but the financial cost was too high to bear for the economy at the time (UNDDA 1991). Although the programme was downplayed in 1960s, both external and internal factors restored it to policy priority in 1970s and 1980s. Externally, the critical roles of the Soviet satellites, and later the US air and space assets, in the intelligence that uncovered a series of South Africa's missile test-flights and the internationally controversial nuclear weapon test-plans made lasting impression, which were difficult to forget and ignore for the apartheid regime (Asuelime and Adekoye 2016; Bidgood 2016; Polakow-Suransky 2010; UNDDA 1991). Moreover, the rise in military capabilities, especially air defence systems and power projection, of Angola, and associated build-up in alliance with the USSR and Cuba among other states, undermined the air power and superiority of South Africa, which is critical to its national defence, security and power projection in the region (Baxter 2012; George 2005; Joubert 2019; Leonard 1983; Minter 1994; Stapleton 2010; IISS 1970–1989). The progress recorded by South Africa's ally, Israel in launching a satellite into orbit also encouraged the regime for this mission (Gottschalk 2010; Polakow-Suransky 2010).

Internal pressure from the Defence Department for locally owned and controlled satellites for C4ISR was salient in South Africa in the 1980s (Albright and Stricker 2016; Hamann 2001; Gottschalk 2010). Although the country was already utilising satellite data purchased from commercial agents like Intelsat, pressure from academic community, space enthusiasts, corporate bodies and government departments also made independent capacity in satellite programme desirable for weather forecast and communication by the apartheid regime in the 1980s (UNDDA 1991; Gottschalk 2010). These among others raised the strategic importance of satellite programme in the country in 1980s and 1990s. Accordingly, the country initiated a Greensat programme, with the plan to launch an Earth Observatory satellite into orbit, between late 1980s and 1990s. This idea evolved from the initial plan to launch a military reconnaissance satellite aboard RSA-3, a solid-fuel orbital launch vehicle, or a solid-propellant RSA-4 (Alberts 2019; Gottschalk 2010; Van Wyk 2014a).

Despite the foregoing initiatives, political and security dynamics in and around South Africa changed drastically, with significant implications for its nuclear, missile and space programmes, in 1990s (Oyewole 2020a, 2023b, c). In 1989, Botha suffered a stroke and resigned as the leader of the National Party (NP) and President of South Africa, and was succeeded by a liberal reformer, Frederik Willem de Klerk, instead of his favourite successor, Berend du Plessis. Faced with internal resistance, the pressure of international anti-apartheid coalition, the global wave of democracy and the receding Cold War threat of Soviet soon after resuming office, de Klerk initiated the process of transition to majoritarian democracy as well as reintegration and repositioning of the country in the international community from 1990 (Berger 2009; Mandela 1995; Sonneborn 2010). The end of Cold War and the collapse of the Soviet Union allayed the fear of communist superpower, which necessitated South Africa's nuclear, and to an extent missile programme and associated deterrence policy. The political transition from apartheid regime to multi-racial and majoritarian democracy in South Africa between 1990 and 1994 also eliminated the core basis for internal and regional resistances, which attracted the intervention of external powers, and undermined the defence and security of the country (Clark and Worger 2011; Dzimba 1998; Jaster 1989; Mandela 1995; Minter 1994; Thompson 2000; Oyewole 2023c).

Due to the foregoing dynamics, the transitional government of South Africa entered into negotiation with relevant actors locally and internationally to display commitment to peaceful co-existence within and among nations, as well as redeem the image of, and clear the sanctions against, the country. In this process, the nuclear programme was terminated in 1990 and weapons were unilaterally destroyed. The country nuclear facilities were opened to international inspection, and the whole denuclearisation was overseen by the International Atomic Energy Agency (IAEA) (Busch and Pilat 2014; Heinonen 2016; Howlett and Simpson 1993; Pabian 2015; Van Wyk 2014b). The transition government equally gave up its medium-range missiles as well as the quest for long-range missiles and space launch vehicles (Gottschalk 2010; Van Wyk 2014a). Amidst these, the country only retained capacity to produce tactical missiles. When the RSA programme was cancelled, the prospect of the satellite that was built in this process was also affected, and the efforts to transform the project into civil and commercial satellite recorded limited progress (Gottschalk 2010). The transitional and post-Apartheid governments signed relevant international treaties to further display their commitments to peaceful use of space and nuclear R&D (Alberts 2019; Grobler 1996; Van Wyk and Van Wyk 2015; Van Wyk 2008, 2009).

The denuclearisation and the general demilitarisation of South Africa, including its space programme by the transitional and post-apartheid government was justified as prudent and a source of soft power for the country (Busch and Pilat 2014; Cock and Mckenzie 1998; le Roux 1999; Onderco and Van Wyk 2019; Pabian 2015; Van Wyk 2012). Amidst these, the huge cost of military expenditure under the apartheid regime cannot be ignored or underestimated. Between 1971 and 1989, South Africa spent over R 63.5 billion (US$ 43.1 billion) on its military. With average of 3.8% of the GDP, when the general government expenditure was 14.6% of the GDP, military expenditure averagely stood at 25.8% of total expenditure in the 19 years (World Bank 2020). The post-Apartheid South African government is more concerned about national integration and development, poverty alleviation, addressing inequality, political stability, peaceful settlement of dispute and international cooperation (Bond 2000; Cawthra 1997; Cock and Mckenzie 1998; le Roux 1999; Edigheji 2010; Fessha 2010; Hanoman 2018; Shapiro and Tebeau 2011). Although the post-apartheid government has spent R 79.3 billion (US$ 65.4 billion) on defence between 1994 and 2018, the military expenditure is averagely 1.36% of the GDP within the 24 years (World Bank 2020).

Despite the demilitarisation of South Africa's space programme and reversal of associated aggressive quest, the country has developed a new road map and course of action in this consideration. Although orbital launcher and satellite programmes were halted, suspended or cancelled during the transition, the post-apartheid government inherited relevant history, institutions, infrastructures and human resources that are required to developed befitting space policy and capabilities. These among other resources have been annexed with varying degrees of success by the post-apartheid government of South Africa to advance its national and human development and security interests. The pressure from of some government departments (as it will be shown later) and the scientific communities locally, as well as international trends in space S&T played major roles in setting the space policy trajectories of post-Apartheid regime in South Africa (Gottschalk 2010; Martinez 2009, 2012a, b, 2016; Munsami 2014; Oyewole 2023a, b; Van Wyk 2009). Geopolitics and the regional hegemonic ambition of the country is also relevant in this policy consideration (Amusan and Oyewole 2017; Froehlich and Siebrits 2019; Ogunnubi and Oyewole 2020; Oyewole 2020a, b). The subsequent section of this chapter examines the policy interests of the post-apartheid government of South Africa.

6.4 Space Policy and Strategy of South Africa

The space policy and strategies of South Africa are products of the politics, history, and socioeconomic development of the country, which are largely presented in the last section. At the same time, they are reflections of the past, present and future efforts and aspirations of the government and people of the country in development and security. Accordingly, this section examines South Africa's space policy objectives in the context of its development and security agendas and extant laws. This section is further subdivided into two, including details on the constitutional and legislative framework, and the strategic policy framework of the subject.

6.4.1 The Constitutional and Legislative Framework of South Africa's Space Policy

The 1996 constitution of South Africa provides some clues on the aspirations of the people and government of the country, which must be promoted in all frontiers, including in space exploration. The Section 7(2) of the constitution mandated the state to "respect, protect, promote and fulfil the rights in the Bill of Rights", which Section 8 (1 and 2) made to be binding on all organs of the state (the legislature, the executive, and the judiciary) and a natural or juristic person within the country (RSA 1996). Accordingly, Sections 9–35 mandated all state and individual actions to respect, protect, promote and fulfil the right of every one—to equality, human dignity, life, privacy, citizenship, security of the person, labour relations, harmless and secure environment, property, political choice, housing, health care, food, water and social security, children, education, language and culture, cultural, religious and linguistic communities, access to information, just administrative action, access to courts as arrested, detained and accused persons, as well as freedom—of religion, belief and opinion, expression, movement and residence, association, assembly, demonstration, picket and petition, trade, occupation and profession, and—from slavery, servitude and forced labour (RSA 1996). These can be said to be a considerable representation of South Africa's national conception and perspective on development and security interests of the people and the government.

Beside the constitution, RSA has a series of space and related legislations that defined the policy and strategic directions of post-apartheid government of the country in engagements with the final frontier of development and security. Earliest among these legislations is the South African Space Affairs Act (SASAA) of 1993. As evident in Section 2(1), the Act has two major objectives. First, to meet all the international commitments and responsibilities of the state in respect to peaceful utilisation of outer space, in order to be recognised as a responsible and trustworthy actor in this consideration. Second, to control and restrict the development, transfer, acquisition, and disposal of dual-purpose technologies, in accordance with the terms of international conventions, treaties and agreements entered into or ratified by the government of the country (RSA 1993). The Act established the South

African Council for Space Affairs (SACSA), which is charged with the responsibility to advice the government, as well as encourage, supervise, implement, licence, and oversee all space activities, in the country. The number and composition of the Council are left for unspecified relevant Minister to decide. Notably, Section 11 barned all forms of launch activities within the RSA, or outside of its territory by registered person in the country, without appropriate licencing (RSA 1993).

In an attempt to satisfy the international pressure and demand on RSA in early 1990s, those that were involved in the conception of SASAA of 1993 arguably underestimated the interests and prospects of post-apartheid regime in space exploration and exploitation, as they largely focus on limiting space and related activities more than promoting them in the country. Accordingly, this Act does not provide the country with coherent space policy or a framework to initiate a space programme. The attempts to address these challenges can be said to have started with pressure from different departments of government and scientific communities, which produced some initiatives taken by the executive arms of government from 2000, as it will be shown later (Martinez 2009; Van Wyk 2009). In December 2005, however, the cabinet approved the proposal to establish SANSA, which was later supported with the legislation of South African National Space Agency (SANSA) Act of 2008. The objective of this Act was clearly stated as follows:

> To provide for the promotion and use of space and co-operation in space-related activities, foster research in space science, advance scientific engineering through human capital, support the creation of an environment conducive to industrial development in space technologies within the framework of national government policy, and for that purpose to establish the SANSA; to provide for the objects and functions of the Agency and for the manner in which it must be managed and governed; and to provide for matters connected therewith (RSA 2008).

The 2008 Act established the SANSA, with the objectives of promoting peaceful utilisation of space; research in space science, communication, and navigation; conducive environment for development of space industry; human capital development and infrastructure in space and advance scientific, engineering and technological competences; and international cooperation in space activities (RSA 2008). The SANSA has the primary responsibility to implement strategy and programme for space activities, which it has to advice the Minister and the government generally on their design, in accordance with adopted national policy. This Act was equally supported by other relevant legislations, including the Astronomy Geography Advantage Act (Act No 21 of 2007) that was assented to in June 2008 as well as the likes of Technology Innovation Agency (TIA) Bill of 2007 (Van Wyk 2009).

Due to the gap between the existing law and need or reality of the country, the South African Department of Trade and Industry (DTI) commissioned a study in 2012 to review SASAA. After a series consultation in 2013 and 2014, SACSA established a six-member Ad Hoc Committee to review the Act. The Committee was later dissolved after Inter-Departmental Working Group was established in 2015 with representatives from DST, DOD, SANSA, Department of International

Relations and Cooperation (DIRCO), Department of Justice and Constitutional Development (DoJ & CD), Department of Telecommunications and Postal Services (DTPS) and Aerospace Maritime & Defence Industries Association (AMD) (Majaja 2015; Majaja et al. 2015; Majaja and Maruping 2017). The Working Group mapped out the areas of inadequacy of the SASAA in its report of the first siting in April 2015, which was presented to the Portfolio Committee, before it had two more meeting in May and September as well as drafting of a new bill in June and public consultation in July/August. As at early 2020, the process of repealing the SASAA remains yet to be completed (SANSA 2020).

As evident in extant laws and relevant strategic policy documents adopted by the country, the government of South Africa put strong emphasis on its international commitments (RSA 1993, 2008; DTI 2009; Alberts 2019; Majaja 2015; Majaja et al. 2015; Majaja and Maruping 2017; Van Wyk 2009). As a result, international conventions and treaties on space signed, ratified, and entered into by the country cannot be ignored in this consideration. Table 6.1 provides a summary of some of these international commitments.

Since 1994, the government of South Africa have adopted, recognised, identified with and supported 11 UN declarations, resolutions and principles on outer space, which are not legally binding but are persuasive and are increasingly considered as international space norms and custom (Alberts 2019; Majaja 2015; Van

Table 6.1 The status of South Africa in international convention and treaties on outer space

Treaties	Details of treaties	Status
Outer Space Treaty	Treaty on Principles Governing the Activities of States in the Exploration and Use of Outer Space, including the Moon and Other Celestial Bodies, entered into force on 10/10/1967	Signed on 01/03/1967 & ratified on 30/09/1968
Rescue Agreement	Agreement on the Rescue of Astronauts, the Return of Astronauts and the Return of Objects launched into Outer Space, entered into force on 3/12/1968	Signed on 06/08/1968 & ratified on 24/09/1969
Liability Convention	Convention on the International Liability for Damage caused by Space Objects, entered into force on 1/9/1972	Signed on 29/03/1972 & ratified on 14/12/2011
Registration Convention	The Convention on Registration of Objects Launched into Outer Space, entered into force on 15/9/1976	Signed and ratified
Nuclear Test Ban Treaty	Treaty Banning Nuclear Weapon Tests in the Atmosphere, in Outer Space and under Water, entered into force on 10/10/1963	Signed & ratified on 10/10/1963
ITSO	Agreement relating to the International Telecommunications Satellite Organisation (ITSO), entered into force on 12/2/1973	Signed on 20/08/1971 & ratified on 12/02/1973

Source Compiled from Albert (2019), Majaja (2015), and Van Wyk (2009)

Wyk 2009). Prominent among these principles are the International cooperation in relation to the peaceful uses of outer space—Resolutions 1721 A and B (XVI) (1961); Declaration of Legal Principles Governing the Activities of States in the Exploration and Use of Outer Space (1963), the Principles Governing the Use by States of Artificial Earth Satellites for International Direct Television Broadcasting (1982), the Principles Relating to Remote Sensing of the Earth from Outer Space (1986), the Principles Relevant to the Use of Nuclear Power Sources in Outer Space (1992) (Alberts, 2019; Van Wyk 2008, 2009; UN 2002). These among others provide the bases for South Africa's space policy and strategy.

6.4.2 The Strategic Policy Framework of South Africa's Space Programmes

Post-apartheid government of South Africa inherited disparate space institutions, initiatives, and programmes, which were begging for policy direction and coordination. Amidst these, the pressure on the policy makers started to yield results for the development of national space policy in early 2000s. A major turning point in this consideration was 2003, when an inter-departmental National Working Group on Space Science and Technology (NWGSST) was established, with representatives drawn from Departments of Communication, Foreign Affairs, Science and Technology, as well as Trade and Industry (Martinez 2009; Van Wyk 2009). Another milestone was recorded in 2005, when the DTI released South African Space Roadmap 2005–2014, with the aim to establish a space agency, develop a space policy and strategy, develop space outreach and capacity building programmes, develop space industry and commercial programme, develop aeronautic and defence strategy, review SASAA, launch a micro-satellite, develop African constellation, among others (Van Wyk 2009). The DTI also commissioned a consortium to develop the National Space Policy Framework in 2006.

In 2009, a 12-page document was published as South Africa's national space policy. The policy statement of the document recognised that the country has several space institutions and programmes that were not well coordinated and managed. Accordingly, guiding policy and institutional frameworks were considered as necessity to streamline the development of capabilities, cooperation and utilisation of outer space for maximum benefits (DTI 2009). The document identified six policy principles, including commitments to the following: (1) the use of outer space for peaceful purposes and benefit of all human; (2) development of robust space capabilities, services and product to support national priorities, coordination and cooperation in governance; (3) responsible behaviours of private and public sectors in space environment in accordance to national legislation and international best practices; (4) promotion of R&D in space science and technology; (5) development of domestic industry to achieve greater national self-reliance in space technology; (6) international cooperation for mutual benefit and peaceful utilisation of space, with emphasis on cooperation with African nations to extend the benefits of space technology to the continent (DTI 2009).

In addition to the six driven principles of the South Africa's space policy, the document also identified eight policy objectives, which tends to restate some of its principles. Accordingly, the government of the country set out to improve coordination of space arena, promote the creation and implementation of relevant regulation to support industrial participation in space arena, foster a robust S&T based higher education and research institutions, promote capacity building initiatives, promote the development of appropriate and competitive commercial space sector to provide the required industrial base for the country in space R&D, promote greater awareness of space S&T and associated benefits at all levels of South African society, provide appropriate and adequate space capabilities to support national and foreign policy objectives, as well as to promote improved mutually benefiting international cooperation in peaceful utilisation of outer space (DTI 2009). Table 6.2 provides a summary of goals and strategies mapped out as general implementation guideline of the space policy.

Beside national space policy, the DST developed and released National Space Strategy, equally referred to as National Space Science and Technology Strategy (NSSTS), which was approved by the Cabinet (SANSA). The undated 19-pages document was suspected to have been published between 2007 and 2009. Among other things, the document covers the strengths and weaknesses, opportunities and threats, and strategic contexts of space activities in South Africa. The highly motivated personnel, industrial capacity as well as the experiences and connectivity of various private corporations, research institutions and universities in space activities are the strengths of the country (DST nd). However, inadequate organisation and coordination of space activities; absence of formalised quality and reliable rules as well as applicable standard in engineering and manufacturing of space products and delivery of related services; and lack of commitments to R&D by private sector are major weaknesses. Amidst these, the level of industrial capabilities for system design and integration; prospects of partnership with MNCs among other international agencies; potential to develop export market from existing facilities and capabilities; and expectation that universities can support industrial development in space sectors are opportunities for the country (DST nd). The only threats that the sector was considered to be faced with is the investment risk of entering a new business domain like space as well as insufficient and emigration of skilled personnel and the challenges of recruiting required foreign specialists in South Africa (DST nd).

The National Space Strategy also covers the strategic context for space activities of the RSA, which include details on the vision, mission, goals, and objectives of the country. In this consideration, the strategic vision is to position the republic among the leading nations in the innovative utilisation of space science and technology to enhance economic growth and sustainable development and thus improve the quality of life for all (DST nd). Moreover, three strategic goals and eleven objectives were mapped out for the country. The following are the three strategic goals in space identified by DST (nd):

Table 6.2 The goals and strategies of South Africa's space policy

No	Strategic goals	Strategic details
1	Development of space capabilities	Addressing national capability requirement in key areas, such as manufacturing, access to space, space science and technology, and human capital
2	Strengthening of space science and technology base	Maximization of the capacities of research and higher institutions to support space activities
3	Development of national space infrastructure	Acquisition and maintenance of various ground facilities, such as facilities required for satellite assembly, testing and integration (ATI), as well as facilities to support flight testing and space launch activities
4	Promotion of domestic space industry	Development of industrial framework, building capacity in the domestic industry, creating a supporting regulatory environment for national space activities, managing innovation and technological transfer to and from the space sector, as well as promoting the competitiveness of space industry
5	Promotion of enhance space awareness at all levels	Sustained and integrated efforts of different government departments, agencies and other space actors to communicate to all levels of the society their activities and their contribution to public goods
6	Enhance international cooperation	Pursue appropriate and mutually beneficial international co-operation opportunities in space R&D according to consistent with its foreign policy objectives (African agenda, and South-South and North–South cooperation) and international obligations; Areas for international co-operation include space science and exploration, Earth observation, communications and positioning, timing and navigation; Active participation in regional and global multilateral fora for the peaceful uses of outer space
7	Funding and investment	Public investment in critical space applications that offers broad societal benefits in the interest of the public; public investment in space infrastructure to create conducive environment for the development of private sector, encourage space tourism and better position the country in the regional and global arenas
8	Monitoring, evaluation and review of the policy	DTI will put mechanism in place to assess the implementation of the policy every ten years, or anytime as deem fit by its Minister

Source Designed by the author with data extracted from DTI (2009)

- To capture a share of global market for small to medium sized space systems in order to support the establishment of a knowledge-based economy through development of innovation and industrial competitiveness.
- To empower the country with better decision-making capacity through the integration of space and ground-based systems that generate accurate and timely information.

- To utilise space science and technology to develop applications that are relevant for the provision of geospatial, telecommunication, timing and positioning products and services.

These goals are further deconstructed in 11 objectives. These are to develop or promote (1) local private space science and technology industry sector; (2) services and products that can respond to user needs; (3) export market for specific equipment for satellite or services offered from existing facilities; (4) existing space S&T activities into strategic programmes; (5) optimum organisation of future space activities to respond to opportunities with international space agencies and industrial partners; (6) partnerships with established and developing countries in space for the purposes of industrial and capacity development; (7) appropriate training and technology transfer programmes, including the sharing of experience and expertise; (8) space S&T in academic institutions and research centres for the provision of opportunities for both short-term and long-term training and education; (9) responsive capacity to challenges and opportunities in Africa; (10) advocacy for space S&T and its importance as a priority measure for meeting national development needs; and (11) capacity to build local awareness of space science and technology and its benefits (DST, nd). In addition to these, the document identified areas of priority for South Africa's space policy, as summarised and presented in Table 6.3.

Table 6.3 Summary of key areas of priority in South Africa's space policy

Key Areas of Priority	Sub-areas of priority
Environmental resource management	• Environmental and geospatial monitoring • Ocean, coastal and marine management • Land management • Rural development and urban planning • Topographic mapping • Hydrological monitoring • Climate change mitigation and adaptation • Meteorological monitoring
Health, safety and security	• Disaster monitoring and relief • Hazard forecasting and early warning • Cross-border risks • Disease surveillance and health risk • Asset monitoring • Regulatory enforcement • Defence, peacekeeping and treaty monitoring
Innovation and economic growth	• Tourism and recreation • Communications • Space science and exploration • Space technology transfer and spin-offs • Development of the space industry

Source Developed by the author with data extracted from DST (nd)

The priority areas are expected to be delivered through space programme, which entail basic areas like earth observation, navigation, communication, and space science and exploration; supporting platforms, which include human capital, infrastructure, and international partnerships; as well as space awareness. Functional programmes for enabling technologies, mission development, space mission operation and space mission applications are considered to be critical in this consideration (DST nd). These among other details on space policy, strategies, and programmes have appeared and have been restated, revised, expanded, and updated in many other official documents. Prominent among these are series of sub-sectional strategic documents, annual reports, strategic plans, and performance plans that are published by the DST and the SANSA (see DST nd2, 3; SANSA 2013/2014; 2015, 2019, 2020).

South African national space policy and strategies cannot be fully understood outside the national development and security priorities that it ought to advance. In this consideration, the government of the RSA adopted a National Research and Development Strategy in August 2002. The strategic document lamented the declining state of R&D in the country, especially in the 1990s. It called for urgent need to address the challenges of R&D, which include appropriate funding of national system of innovation, human resources, areas of strategic considerations, declining R&D in private sector, intellectual property questions, as well as fragmentation of governmental measures on science and technology (Government of the RSA 2002). Accordingly, the document set the following strategic objectives and priorities for the country: to achieve mastery of technological innovation and change in society and economy, increase investment in science based human capital and national transformation, as well as to create effective government institutions that can align and deliver science and technology system (Government of the RSA 2002). Amidst these, the document acknowledged the progress made by the USA, the former USSR, and European countries in space programme among other R&D initiatives in S&T, and the implications for their various societies (Government of the RSA 2002).

The South African Defence Review of 2015 equally emphasise the developmental nature of the state and the need to protect it. The review identified four strategic gaols and 13 tasks for national defence of the country. The goals are to defend and protect the RSA, safeguard the RSA, promote peace and security, as well as developmental and other ordered tasks (DOD 2015). To defend and protect South Africa, the review prioritised deterrence and prevention of conflict, protection of national interests, and defence of the country. To safeguard South Africa, the tasks are to safeguard the borders and critical infrastructures, cooperate with the police services and ensure information security. Promotion of strategic influence as well as contribution to peace and stability are essentials to achieve goal three: promotion of peace and stability, especially internationally. Developmental and other ordered tasks were deconstructed in the review as execution of relevant treaty obligations, performance of ordered presidential tasks, assist civil authority as ordered as well as contribution to the development of South Africa and its people (DOD 2015). Although without much emphasis, DOD (2015) recognised the relevance of space assets and programme amidst air, land, maritime and cyberspace assets to achieve the desired defence and security objectives of the country.

6.5 Conclusion

This chapter provides some insights on South Africa's space policy and strategies, their place in the country's conceptions and quest for development and security, as well as how they are shaped by political, historical, and socioeconomic trajectories. Amidst these, it can be observed that the South African conceptions of development reflect some of the conceptual perspectives detailed in chapter two. The ends of development of the RSA reflect economic growth, transformation, catching up, industrialisation, liberalisation and privatisation, resource management, socio-political modernisation, national capacity, human and community flourishing, enlargement of people's choice as well as sustainability. The conceptions of security in the country also cover elements of national and human dimensions of security, as earlier detailed in chapter two. The conception of space support for these development and security goals aligns with utilitarian framework of analysis adopted in this study. This chapter, therefore, provides the basis for the next chapters on the development of space capacities and capabilities in South Africa, and their contribution to development and security aspiration of the people and government of the country.

References

AfDB et al. 2019. *African statistics yearbook 2019*. Abidjan and Addis Ababa: African Development Bank Group (AfDB) the African Union Commission (AUC) and the United Nations Economic Commission for Africa (ECA).

Alberts, A. 2019. South Africa. In *Integrated space for African society: Legal and policy implementation of space in African countries*, ed. A. Froehlich. Switzerland: Springer.

Albright, D., and A. Stricker. 2016. *Revisiting South Africa's nuclear weapons program: Its history, dismantlement, and lessons for today*. Washington, DC: Institute for Science and International Security (ISIS) Press.

Amusan, L. 2016. Spoils and meritocracy: Post-apartheid challenges in the South African Public service. *Journal of Public Administration* 51 (2): 293–305.

Amusan, L., and S. Mchunu. 2017. An assessment of xenophobic/afrophobic attacks in South Africa (2008–2015): Whither Batho Pele and ubuntu principles? *South African Review of Sociology* 48 (4): 1–18.

Amusan, L., and S. Oyewole. 2017. The quest for hegemony and the future of African solution to African development problems: Lessons from headways in African security sector. *Journal of Asian and African Studies* 52 (1): 21–33.

Anderson, N., and M.S. Bell. 2019. The limits of regional power: South Africa's security strategy, 1975–1989, *Journal of Strategic Studies*. https://doi.org/10.1080/01402390.2019.1695123.

Asuelime, L.E., and R.A. Adekoye. 2016. *Nuclear proliferation in South Africa: History and politics*. Switzerland: Springer.

Baxter, P. 2012. *SAAF's border war: The South Africa's air power in combat, 1966–1989*. England: Helion & Company Limited

Beinart, W., and S. Dubow, eds. 1995. *Segregation and apartheid in twentieth-century South Africa.* London and New York: Routledge.
Berger, I. 2009. *South Africa in world history.* Oxford: Oxford University Press.
Bidgood, S. 2016. The 1977 South Africa nuclear crisis. *Adelphi Series* 56 (464–465): 55–78.
Bin, L. 1997. Nuclear missile delivery capabilities in emerging nuclear states. *Science & Global Security: THe Technical Basis for Arms Control, Disarmament, and Nonproliferation Initiatives* 6 (3): 311–331.
Bond, P. 2000. *Elite transition: From apartheid to neoliberalism in South Africa.* London: Pluto Press.
Burgess, S.F. 2006. South Africa's nuclear weapons policies. *The Nonproliferation Review* 13 (3): 519–526.
Busch, N.E., and J.F. Pilat. 2014. South African rollback: Revisiting monitoring and verification lessons after 20 years. *Comparative Strategy* 33 (3): 236–261.
Cawthra, G. 1997. *Securing South Africa's democracy: Defence, development, and security in transition.* London: Macmillan Press Ltd.
CIA. 2020. World fact book: Africa: South Africa. https://www.cia.gov/library/publications/the-world-factbook/geos/sf.html.
Chun, CKS. 2006. *Thunder over the horizon: From V-2 rockets to ballistic missiles.* Westport: Greenwood Publishing Group, Inc.
Cillie, G. G. 1973. Brief survey of the development of astronomy in South Africa. *Monthly Notes of the Astron. Soc. Southern Africa,* (32): 59–.
Clark, N.L., and W.H. Worger. 2011. *South Africa: The rise and fall of apartheid*—, 2nd ed. London and New York: Routledge.
Cochran, T.B. 1994. Highly enriched uranium production for South African nuclear weapons. *Science & Global Security: THe Technical Basis for Arms Control, Disarmament, and Nonproliferation Initiatives* 4 (2): 161–176.
Cock, J., and P. Mckenzie. 1998. *From defence to development: Redirecting military resources in South Africa.* Claremont: David Philip Publishers (Pty) Ltd.
Department of Trade and Industry (DTI). 2008. *National space policy.* Pretoria: Government of the Republic of South Africa.
DOD. 2015. *Annual report 2014/2015.* Pretoria: DOD.
DST. Nd. *National space strategy: To leverage the benefits of space science and technology for socio-economic growth and sustainable development.* Pretoria: DST.
DST. Nd2. *DST ten years innovation plan.* Pretoria: DST.
DST. Nd3. *South African earth observation strategy.* Pretoria: DST.
DTI. 2009. *South Africa's national space policy.* Pretoria: DTI.
Dubow, S. 2019. 200 years of astronomy in South Africa: From the royal observatory to the 'Big Bang' of the square kilometre array. *Journal of Southern African Studies* 45 (4): 663–687.
Dzimba, J. 1998. *South Africa's destabilization of Zimbabwe, 1980–89.* London: Macmillan Press Ltd.
Edigheji, O, ed. 2010. *Constructing a democratic developmental state in South Africa: Potentials and Challenges.* Cape Town: Human Science Research Council Press.
Evans, D.S. 1967. Historical notes on astronomy in South Africa. *Vistas in Astronomy* 9: 265–282.
Evans, D. S. 1960. Some aspects of South African astronomy. *Annales d'Astrophysique* 23: 436–.
Feinstein, C.H. 2005. *An economic history of South Africa: Conquest, discrimination and development.* Cambridge: Cambridge University Press.
Fessha, Y.T. 2010. *Ethnic diversity and federalism: Constitution making in South Africa and Ethiopia.* Farnham and Burlington: Ashgate.
Froehlich, A., and A. Siebrits. 2019. *Space supporting Africa: A primary needs approach and Africa's emerging space middle powers.* Cham: Springer.
Gastrow, M. 2015. *The stars in our eyes: Representations of the square kilometre array telescope in the South African Media.* PhD diss.: University of Stellenbosch, South Africa.
Gastrow, M., and T. Oppelt. 2019. The square kilometre array and local development mandates in the Karoo. *Journal of Southern African Studies* 45 (4): 711–728.

Gastrow, M., G. Kruss, and I. Petersen. 2016. Connecting capabilities in highly unequal developing countries: The case of the Square Kilometre Array telescope in South Africa. *Development Southern Africa* 33 (3): 361–375.

GCIS. 2019. *South Africa yearbook 2018/19* (26th edition). Pretoria: Government Communications (GCIS)

George, E. 2005. *The Cuban intervention in Angola, 1965–1991: from Che Guevara to Cuito Cuanavale*. London and New York: Frank Cass.

Goodson, D.L.R. 2012. Catalytic deterrence? Apartheid South Africa's nuclear weapons strategy. *Politikon* 39 (2): 209–230.

Gottschalk, K. 2010. South Africa's space program. *Astropolitics* 8 (1): 35–48.

Government of the RSA. 2002. *South Africa's national research and development strategy*. Pretoria: The Government of the Republic of South Africa.

Grab, S., and J. Knight. 2015. *Landscapes and landforms of South Africa*. New York: Springer.

Grobler, G.J. 1996. The South African experience: The birth and death of South Africa's nuclear weapons programme. *Medicine and War* 12 (3): 228–239.

Guelke, A. 2005. *Rethinking the rise and fall of apartheid: South Africa and world politics*. New York: Palgrave Macmillan.

Hamann, H. 2001. *Days of the generals: The untold story of South Africa's apartheid-era generals*. Cape Town: Zebra Press.

Hanoman, J. 2018. *Hunger and poverty in South Africa: The hidden faces of food insecurity*. London and New York: Routledge.

Harris, V., S. Hatang, and P. Peter Liberman. 2004. Unveiling South Africa's nuclear past. *Journal of Southern African Studies* 30 (3): 457–476.

Harvey, B., H.H.F. Smid, and T. Pirard. 2010. *Emerging space powers: The new space programs of Asia, the Middle East and South America*. New York: Springer.

Heinonen, O. 2016. Lessons learned from dismantlement of South Africa's biological, chemical, and nuclear weapons programs. *The Nonproliferation Review* 23 (1–2): 147–162.

Holbrook, J., Medupe, R.T., and J.O. Urama, eds. 2008. *African Cultural Astronomy: Current Archaeoastronomy and Ethnoastronomy Research in Africa*. Springer.

Holbrook, J. 2015. Cultural Astronomy in Africa South of the Sahara. In Ruggles, C.L.N. (ed.), *Handbook of Archaeoastronomy and Ethnoastronomy*. New York: Springer.

Howlett, D., and J. Simpson. 1993. Nuclearisation and denuclearisation in South Africa. *Survival*, 35 (3): 154–173.

IISS (Institute of International Security Studies). 1970–1989. *The military balance*. London: IISS.

Jaster, R.S. 1989. *The defence of white power: South Africa's foreign policy under pressure*. New York: Palgrave Macmillan.

Joubert, S. 2019. Gunship over angola: The story of a maverick pilot. Johannesburg & Cape Town: Delta Books. IISS (Institute of International Security Studies). The Military Balance, 1970–1989. London: IISS.

Landau, P.S. 2010. *Popular politics in the history of South Africa, 1400–1948*. Cambridge: Cambridge University Press.

Lastovica, E. 1995. Ardour in the cause of astronomy: Bibliography of the publications of Sir Thomas Maclear with an historical introduction. *Transactions of the Royal Society of South Africa* 50 (1): 65–77.

le Roux, L. 1999. Defining defence requirements: Force design considerations for the South African National Defence Force. *African Security Studies* 8 (5): 56–64.

Leonard, R. 1983. *South Africa at war: White power and the crisis in Southern Africa*. Westport and Connecticut: Lawrence Hill & Company.

Liberman, P. 2004. Israel and the South African bomb. *The Nonproliferation Review* 11 (2): 46–80.

Livsey, T. 2017. *Nigeria's university age: Reframing decolonisation and development*. London: Palgrave Macmillan.

Lodge, T. 1989. *Black politics in South Africa since 1945 (6th impression)*. New York: Longman Inc.

Majaja, N., and P. Maruping. 2017. Repeal of the space affairs act no. 84 of 1993 by the South African outerspace bill. Portfolio Committee on Trade and Industry, 20 June 2017.

Majaja, N., P. P. Sekhula, and P. Martinez. 2015. Review of the space affairs act no. 84 of 1993. Briefing for the Parliamentary Portfolio Committees of Trade & Industry and Science & Technology, May 12.

Majaja, N. 2015. Review of the space affairs act no. 84 of 1993. Parliamentary Portfolio Committee: Trade and industry, February 24.

Mandela, N. 1995. *Long walk to freedom: The autobiography of Nelson Mandela.* New York: Back Bay Books.

Martinez, P. 2009. Building capacity in the basic space sciences in Southern Africa: Experiences and prospects. *Advances in Space Research* 43: 1866–1872.

Martinez, P. 2012b. Is there a need for an African space agency? *Space Policy* 28: 142–145.

Martinez, P. 2016. The development and initial implementation of South Africa's national space policy. *Space Policy* 37: 30–38.

Martinez, P. 2012a. The African leadership conference on space science and technology for sustainable development. *Space Policy* 28(2012): 33–37.

Masiza, Z. 1993. A chronology of South Africa's nuclear program. *The Nonproliferation Review/ Fall*, 34–54.

Mathenjwa, L. F. 1999. *The Zulu literary artist's conception of celestial bodies and associated natural phenomena.* Doctoral Thesis of Literature and Philosophy in African Languages, University of South Africa.

Minter, W. 1994. *Apartheid's contras: An inquiry into the roots of war in Angola and Mozambique.* London and New Jersey: Zed Books.

Moore, J.D.L. 1987. *South Africa and nuclear proliferation: South Africa's nuclear capabilities and intentions in the context of international non-proliferation policies.* New York: St. Martin's Press Inc.

Munsami, V. 2014. South Africa's national space policy: The dawn of a new space era. *Space Policy* 30: 1–6.

Ogunnubi, O., and S. Oyewole, eds. 2020. *Power politics in Africa: Nigeria and South Africa in comparative perspective.* Newcastle: Cambridge Scholar Publishing.

Onderco, M., and A.M. Van Wyk. 2019. Birth of a norm champion: How South Africa came to support the NPT's indefinite extension. *The Nonproliferation Review* 26 (1–2): 23–41.

Oyewole, S. 2017. Space research and development in Africa. *Astropolitics* 15 (2): 185–208.

Oyewole, S. 2020a. The quest for space capabilities and military security in Africa. *South African Journal of International Affairs* 27 (2): 147–172.

Oyewole, S. 2023a. The African orbital journey in space age: From spectator to participant. *Vanguardia Dossier* 88 (15): 72–75.

Oyewole, S. 2023c. Defence and security policy of South Africa. In *Readings in defence, security and strategic studies*, ed. F.C. Onuoha, G.E. Ezirim, and C.M.A. Kwaja, 124–135. Nsukka: University of Nigeria Press Limited.

Oyewole, S. 2020b. Military capabilities of regional powers in Africa: Nigeria and South Africa in comparative perspective. In *Power politics in Africa: Nigeria and South Africa in comparative perspective*, ed. Olusola Ogunnubi and Samuel Oyewole. Newcastle: Cambridge Scholar Publishing Ltd.

Oyewole, S. 2023b. Military space strategies and African realism: Egypt, South Africa and Nigeria. In *Oxford handbook of space security*, ed. S. M. Pekkanen and P. J. Blount. Oxford: Oxford University Press.

Pabian, F.P. 2015. The South African denuclearization exemplar. *The Nonproliferation Review* 22 (1): 27–52.

Pillay, S. 2014. *Development corruption in South Africa: Governance matters.* New York: Palgrave Macmillan.

Polakow-Suransky, S.P. 2010. *The unspoken alliance: Israeli's secrete relationship with apartheid South Africa.* New York: Pantheon Books.

Purkitt, H.E., and S.F. Burgess. 2005. *South Africa's weapons of mass destruction.* Bloomington: Indiana University Press.

RSA. 1993. Space Affairs Act (South Africa, 1993). Statutes of the Republic of South Africa—Trade and Industry No. 84 of 1993.

RSA. 1996. *The constitution of the Republic of South Africa.* RSA.

RSA. 2008. *South African national space agency act.* Cape Town: RSA.

SANSA. 2013/2014. *SANSA annual performance plan 2013/2014.* Pretoria: SANSA.

SANSA. 2015. *SANSA annual performance plan 2015–2020.* Pretoria: SANSA.

SANSA. 2019. *SANSA annual report 2018/2019.* Pretoria: SANSA.

SANSA. 2020. *SANSA annual performance plan 2020/2021.* Pretoria: SANSA.

Schulz-Herzenberg, C. 2007. *A lethal cocktail: Exploring the impact of corruption on HIV/Aids prevention and treatment efforts in South Africa.* Pretoria and Harari: Institute for Security Studies and Transparency International

Seekings, J., and N. Nattrass. 2005. *Class, race, and inequality in South Africa.* New Haven and London: Yale University Press.

Shapiro, I., and K. Tebeau. 2011. *After apartheid: Reinventing South Africa.* Charlottesville: University of Virginia Press.

Smedegar, K. 2007. Problems and prospects in the cultural history of South African astronomy. *African Skies* 11: 27–32.

Sonneborn, L. 2010. *The end of apartheid in South Africa.* New York: Infobase Publishing.

Space Affairs Act (South Africa, 1993). In *National legislation and policy: Selected texts.* United Nations/Nigeria Workshop on Space Law Hosted by the Government of Nigeria, Abuja, 21–24 Nov 2005.

Stapleton, T.J. 2010. *A military history of South Africa: From the Dutch-Khoi wars to the end of apartheid.* Santa Barbara: Praeger.

Stephens, J. J. 2003. *Fuelling the empire: South Africa's gold and the road to war.* Wester Sussex: John Wiley and Son Ltd.

Tafira, H. K. 2018. *Xenophobia in South Africa: A history.* Palgrave Macmillan.

Thompson, L. 2000. *A history of South Africa,* 3rd ed. New Haven: Yale University Press.

Thomson, A. 2008. *U.S. foreign policy towards apartheid South Africa, 1948–1994: conflict of interests.* New York: Palgrave Macmillan.

Twidle, H. 2019. Impossible images: Radio astronomy, the square kilometre array and the art of seeing. *Journal of Southern African Studies* 45 (4): 767–790.

UN. 2002. *United nations treaties and principles on outer space.* New York: United Nations Publications

UNDDA (United Nations Department of Disarmament Affairs). 1991. *South Africa's nuclear-tipped ballistic missile capability.* New York: United Nations Publications.

Urama, J. O., and J. C. Holbrook. 2011. The African cultural astronomy project. In *The role of astronomy in society and culture,* ed. D. Valls-Gabaud and A. Boksenberg. Proceedings IAU Symposium No. 260, 2009.

Urama, E. N. 2008. *Cultural astronomy in African literature. master thesis of arts in literary arts.* Department of English and Literary Studies, University of Nigeria, Nsukka.

Van Wyk, M.S. 2007. 'Ally or critic? The united states' response to South African nuclear development, 1949–1980'. *Cold War History* 7 (2): 195–225.

Van Wyk, A. 2010. Apartheid's atomic bomb: Cold war perspectives. *South African Historical Journal* 62 (1): 100–120.

Van Wyk, J.A. 2012. Nuclear diplomacy as niche diplomacy: South Africa's post-apartheid relations with the international atomic energy agency. *South African Journal of International Affairs* 19 (2): 179–200.

Van Wyk, J.A. 2014b. Atoms, apartheid, and the agency: South Africa's relations with the IAEA, 1957–1995. *Cold War History.* https://doi.org/10.1080/14682745.2014.897697.

Van Wyk, J.A., and M.S. Van Wyk. 2015. From the nuclear laager to the non-proliferation club: South Africa and the NPT. *South African Historical Journal* 67 (1): 32–46.

van Wyk, J.A. 2008. Overview of the implementation status of the five united nations treaties on outer space in African countries. *African Skies* 12: 20–28.

van Wyk, J.A. 2009. South Africa's space policy and interests: A new dawn or a black hole? *Strategic Review for Southern Africa* 31 (2): 46–73.

van Wyk, A. 2018. South African nuclear development in the 1970s: A non-proliferation conundrum?". *The International History Review*. https://doi.org/10.1080/07075332.2018.1428212.

Walker, C., and D. Chinigò. 2018. Disassembling the square kilometre array: Astronomy and development in South Africa. *Third World Quarterly*, https://doi.org/10.1080/01436597.2018.1447374.

Warner, B. 2008. Astrophysics in South Africa. *Transactions of the Royal Society of South Africa* 63 (1): 61–67.

Warwick, R. 2009. *White South Africa and defence 1960–1968: militarization, threat perceptions and counter strategies*. PhD Thesis, Department of Historical Studies, University of Cape Town, South Africa.

World Bank. 2020. World development indicators. Available at http://datatopics.worldbank.org/world-development-indicators/. Accessed 10 March 2021.

Van Riper, A.B. 2004. Rockets and missiles: The life story of a technology. In *Nuclearisation and denuclearisation in South Africa*, Survival, ed. D. Howlett, and J. Simpson. 1993, 35(3): 154–173. Westport: Greenwood Press.

van Wyk, J. A. 2014a. Apartheid South Africa's nuclear weapons programme and its impact on Southern Africa. *Austral: Brazilian Journal of Strategy & International Relations* 3(6): 119–140.

The Development of Space Capacities and Capabilities in South Africa

7

Abstract

This chapter examines various South African institutions (governmental, non-governmental and private entities) that are involved in space policy processes, the resources that are available to them (such as funds, human and materials: facilities and equipment), and their capacities to mobilise and covert these into capabilities (such as space assets and access). Beyond different government entities for civil, commercial and military space programmes and NGOs, South Africa has the most deregulated African space industry with flourishing private sector. Building on its space heritage, the government has supported science and technological innovations in its academic institutions, which have generated spin-off companies that are playing critical roles in the realisation of set objectives in the national space policy and strategy. This understanding of space capacities and capabilities of South Africa is important to the nexus between the national interests in space, and the extent that such policy objectives have been realised, especially in advancing the development and security aspirations of the people and government.

7.1 Introduction

This chapter examines the institutions, relevant resources, their conversion and mobilisation for South Africa's space policy, and how they are operationalised for space access and assets of the country. As earlier stated, starting a space programme and progress in development of relevant capacities and capabilities have become a form of developmental goal globally since the dawn of space age. Accordingly, establishing a space agency, buying, developing, owing, orbiting and operating a satellite, publishing space policy document and strategic paper, developing and operating a satellite ground stations, observatories and other facilities, developing and operating orbital vehicle are development objectives of many

S. Oyewole, *Utilitarianism in Outer Space*, Southern Space Studies,
https://doi.org/10.1007/978-3-031-49646-2_7

nations in space age (Froehlich and Siebrits 2019; Harding 2013; Mistry and Gopalaswamy 2012; Wood and Weigel 2011; Oyewole 2017). This is the case of many developing countries and emerging powers, such as South Africa (Luncedo and Gottschalk 2013; Martinez 2016; Munsami 2014; Oyewole 2023a, b).

As evident in previous chapter, the policy and strategic goals of South Africa in development and security, in relation to space frontier, include development of necessary institutional, human resources, industrial and technological capacities and capabilities to fully realise the potential of the country and maximise accruable benefits from space and the attendant technologies. Consequently, the progress that have been made by the country in space programming and capacity building, and capabilities cannot be ignored in the assessment of its space policy and strategy. It will be impossible to fully appreciate the contribution of space policy and strategy to the development and security in South Africa (output) without adequate knowledge of the resources that were committed and capabilities that were developed for this purpose (input).

This chapter examines the institutions, activities, assets, human resources, funding, and skills among other resources that are made available to operationalise the space policy aspirations of the people and government of South Africa. To achieve this objective, this chapter is divided into three broad sections in addition to this introduction and the concluding section. The first section assesses the state and non-state institutions that have been developed for civil, commercial and defence space policy and strategies of the country. The second section examines the financial, human and material resources that are made available and mobilised to operationalise institutions that are relevant to the attainment of the space policy objectives of South Africa. The third section examines how these institutions and resources are responsible and reflected in South Africa's space access and assets. These are followed by a concluding section.

7.2 Institutional Development for Space Policy and Programmes in South Africa

Institution is an important building block and component of development, including in space R&D. Space policy and strategies are designed and operationalised by relevant institutions, which are expected to transform sporadic, unintended, ad hoc and less or un-coordinated interests and activities into coherent and substantive programme to maximise the utility derived from available resources. As indicated in the previous chapter, a nation's space institution broadly covers publicly owned (by the state/governmental and non-state/governmental) entities and private organisations. NGOs and private sector are increasingly playing critical roles in national space policy globally, although governmental institutions remain central to create and maintain required environment for them to thrive as well as coordinate their operations in line with the national interests. It is against this background that this section is divided into two sub-sections. The first one examines the development of governmental space institutions, while the other surveys the development of space-related NGOs and private actors, in South Africa.

7.2.1 Development of State Institutions for Space Policy and Programmes in South Africa

South Africa can be said to run three broad categories of space programme, including civil, commercial, and military space programmes. In some of these cases, the republic has institutionalised these programmes, as a number of institutions have been developed over the years for these purposes. Table 7.1 provides a summary of government space programmes and the major institutions developed for these purposes in the RSA. Accordingly, this section is further divided into three sub-sections. The first sub-section examines government departments and agencies, as well as some of the collaborating academic institutions locally, which are central to operationalise the civil space programme. The second sub-section deals with the state owned and sponsored commercial entities that are involved in space affairs, while the third sub-section provides insights into the state institutions for military space programme in the RSA.

7.2.1.1 Development of State Institutions for Civil Space Programmes in South Africa

As identified in the last chapter, modern space S&T in South Africa can be traced to the 1700s. Accordingly, many institutions have emerged for this purpose over the centuries. However, it was only in early 2000s that concrete steps were taking to coordinate the disparate institutions and their activities. This gave birth to the establishment of SANSA in December 2010. The politics behind this development has been considered in the last chapter. As a consequence, this section is only interested in the organisation, its functionalities and link to other institutions.

Unlike the Nigerian space legislation that created a decentralised space agency with multiple centres, the South African space legislation created a relatively centralise space agency. Although the later recognised the need for the space agency to create or partner with a company as well as establish, coordinate and manage

Table 7.1 State-owned space institutions in South Africa

Space programmes	Institutions and centres
Civil space institutions	SANSA + relevant Departments (DST, DHET, DTI & DST) and agencies/institutions (CSIR, SACSA, & NRF) Major collaborating academic institutions locally: CPUT, UCT, UWC, University of Stellenbosch, TUT, UNISA, UKZN, University of Free State, University of Pretoria, & Wits
Commercial space institutions	Spaceteq, Sentech Soc Ltd
Military space institutions	DOD: Armscor & SANDF (SAAF, SAA, SAN, DI); & SSA

Source Compiled by the author

any programme considered necessary, SANSA is made the central institutional unit of space policy implementation in South Africa (RSA 2008). In the case of Nigeria, NASDRA has several centres that are created by law, funded directly from the National Appropriation, and exercise some levels of administrative and operational autonomy from the central institution (see chapter four for more details). In the RSA, however, the SANSA has facilities that operate like departments with limited institutional autonomy, as they are managed centrally or remotely, in case of distance ones, and by partnered organisations.

The operations of SANSA cover four major programmes, which include EO, space engineering, space operations, and space science. The EO programme entails collection, processing, archiving and dissemination of various EO satellite data to support decision-making, economic growth, and sustainable development in the RSA. The space engineering programme is designed to develop, build, and test satellite systems and sub-systems. The space operation programme focuses on various ground station facilities and services for launch activities, including satellite tracking, telemetry, and command (TT&C), launch support, in-orbit testing, mission control and space navigation. Space Science operates programme for the study of the Earth's magnetic field, the Sun, and the near-space environment (SANSA 2020b). The SANSA head office is located in Pretoria, from where the EO and space engineering programmes are also coordinated. The SANSA space operations programme is coordinated from Hartebeesthoek, while the space science is coordinated from Hermanus in Western Cape (SANSA 2020c).

Although SANSA is the primary state institution in the implementation of space policy in South Africa, it operates closely and collaboratively with several government departments, agencies, and institutions. The agency maintains close working relationships with DST and other departments, such as the Department of Higher Education and Training (DHET), the Department of Communications and Digital Technologies (DCDT), DTI, DOD, and the DIRCO. As an agency under DST, which is also known as the Department of Science and Innovation (DSI), SANSA operates closely with the central administration of the Department as well as associated agencies or related institutions within it. Prominent among these institutions are the Academy of Science of South Africa, the CSIR, the National Advisory Council on Innovation (NACI), the National Research Foundation (NRF), and the TIA.

The Academy of Science of South Africa was established by President Mandela in May 1996. The Academy currently operates within the legislative framework of Academy of Science of South Africa Act (Act 67 of 2001), which was passed and amended by the Parliament, and came into operation in May 2002. The Academy has a physical presence in Pretoria. The mandate of the organisation entails using science to investigate national problems and provide evidence-based solutions; inspire pure and applied sciences for the benefit of the society; connect the country to the highest level of knowledge globally; and facilitate public understanding of the character, scope and value of scientific and technological enterprise (Academy of Science 2020). The focus of the Academy covers various fields of science, including space science.

The CSIR is another relevant player in space R&D in RSA. The CSIR was established by Act of Parliament in 1945. Its office is based in Pretoria. The mandate of the organisation, as stated by the Scientific Research Council Act 46 of 1988, amended by Act 71 of 1999, is to promote and direct multi-disciplinary research and technology for the advancement national interests. It is also mandated to work with private and public sectors to promote critical fields of science, industrial research, and development to improve the quality of life of the people of the RSA (CSIR 2020). As evident in the last chapter, this organisation played important role in the formulation and implementation of space policy of the country, especially before 2000s. Thereafter, the activities of the CSIR have directly or indirectly supported space R&D in South Africa.

The NACI is another state-owned organisation that is relevant for space R&D in South Africa. The organisation operates with the legislative framework of The National Council on Innovation Act, 1997 (Act No. 55 of 1997). It has the mandate to promote STEM research and studies to improve and sustain quality of life, develop human capital for science and technology, and building the economy of South Africa, as well as strengthen the competitiveness of the country in the international system (NACI 2020). This makes it possible for the activities of NACI to support space R&D directly or indirectly in the RSA. This organisation maintains a physical presence in Pretoria.

The NRF is another government establishment that is involved in promotion of science and technology R&D. Following a major review of the Department of Art, Culture, Science and Technology (DACST), NRF was established in 1998 to take over the functions and services of the pre-existing research funding entities, such as the Centre for Science Development (CSD) that was dedicated to Human Sciences Research Council (HSRC) as well as the Foundation for Research Development (FRD) and its affiliated research facilities (NRF 2020a). The entity is mandated by the National Research Foundation Act (Act No 23 of 1998) to fund advance research, promote human capacity development, as well as provide, support and maintain research facilities in all fields of science and technology, without excluding humanities, social sciences and indigenous knowledge. The foundation also has the mandate to bridge the gap between the scientific community and the society through public awareness, as well as support government and its policy priorities with scientific R&D (NRF 2020b). The NRF maintain a physical presence in Pretoria. As an entity under the DST, the foundation work closely with other agencies in the department and Higher Education Institutions (HEIs) across the country.

The TIA is another entity in the DST that is relevant to technological R&D, including in the space sector. The Agency was established and operates within the legislative framework of Technology Innovation Act (Act 26 of 2008). The mandate of the Agency is to stimulate and promote innovative R&D, discoveries, and exploitation in public interest. For these purposes, the Agency maintain physical presence in Pretoria, Cape Town, and Durban. The TIA has developed pre- and post-investment initiatives to support technological research and development in South Africa from proof of conception to pre-commercialisation. These among other things led the agency to establish various funds, which include the seed fund, the technology development fund, and the commercialisation support fund (TIA 2020).

Beyond the DST, the DHET and affiliated higher institutions are state entities that have played critical roles in the formulation and implementation of South Africa's space agency. The DHET was established in 2009, when the former Department of Education (DE) was divided into two new ones: Department of Basic Education (DBC) and the DHET. The Department has the mandate to undertake policy measures to provide, support and promote education, training, and skill development beyond basic schooling. It operates under the constitutional provision of Section 29 on educational rights of all as well as the legislative framework of the Higher Education Act, the Continuing Education and Training Act, 2006 (Act No 16 of 2006) (SET Act), the Further Education and Training Act (AET Act and FET Act), the National Student Financial Aid Scheme (NSFAS Act), the National Qualification Framework Act (NQF Act), and the Skill Development Levies Act (DHET 2020). The Department develop the policy framework, accredit, and supervise universities in the RSA.

South African universities played important roles in space policy processes, most especially the implementation and assessment, in the country. The government of RSA and the space agency have engaged institutions, experts, and students at the University of Stellenbosch, UCT and CPUT. Other major universities in this consideration are the Tshwane University of Technology (TUT), the University of Western Cape (UWC), the University of KwaZulu Natal (UKZN), the Nelson Mandela Metropolitan University (NMMU), the University of Free State, the University of Pretoria, the University of Witwatersrand (Wits) and the UNISA. The Faculties of Science and Engineering of these universities and their products among others are playing critical roles in managing many space facilities and corporations, as well as catalysts of innovations and inventions in this sector of the country. Amidst these, University of Stellenbosch, UCT and CPUT have made some of the most important landmarks in space policy implementation of the country, considering their instrumentation in satellite construction. Notably, the French South African Institute of Technology (F'SATI), which is jointly developed by the French and South African governments, is based in both the CPUT and TUT. Similarly, the UCT maintains a Spacelab and the University of Stellenbosch has an Electronic System Laboratory (ESL), which have played major roles in South Africa's space sector as it will be shown later in this chapter.

The SANSA also work closely with the DCDT in implementation of South Africa's space agency. The DCDT was created in June 2019, following the merger of the Department of Communication (DOC) and the DTPS. Among other things, the Department has the responsibility to promote, coordinate and manage interest and investment in Information and Communication Technologies (ICT), and ensure that they are utilised to enhance the socio-economic development of the people of South Africa (National Government 2020). The mandate of the department includes both terrestrial and satellite communications, as well as ICT support for other forms of space assets. These, among other, gave the DCDT some implicit and explicit roles in the formulation and implementation of space policy in the RSA.

The DTI and DoD are also major institutions in the commercial and defence dimension of space policy in South Africa, and they will be considered in the appropriate sections later in this chapter. The relevance of space assets and facilities for the activities of some other departments of government also make them institutionally important for the implementation of South Africa's space policy and strategy, especially in utilitarian of space technologies. Prominent among these departments are the Department of Environment, Forestry and Fisheries (DEFF), the Department Agriculture, Land Reform and Rural Development (DALRRD) and the Department of Transportation (DOT) as well as their various agencies and institutions that relied on space assets and facilities to support their policy processes and operations.

It is against this background that SANSA and its associated structure cannot be taken out of context. These and other government departments and their agencies as well as most of the affiliated research and academic institutions are critical state-institutions in the South African civil space programme. Accordingly, the state institutional development for the civil space programmes of South Africa cannot be reduced to SANSA, even though the organisation remains central in the whole process. Instead, the subject of state institutional development for civil space policy and programmes in the country can be approached from two perspectives. One of these approaches is focus on SANSA as the central unity of analysis, while the other tends to recognise network of many other entities, in state institutional development for civil space programmes in South Africa.

7.2.1.2 Development of State Institutions for Commercial Space Programmes in South Africa

The development of state-owned commercial space entities in South Africa is an offshoot of the country's civil and military space programmes. It is important to note that these entities are to various degrees subjected to the regulatory, supervisory, partnership and oversight roles of the SANSA, DST, DOD, DCDT and DTI. Prominent among the state institutions developed for commercial purposes in the space sector in the RSA are Spaceteq and Senntech Soc Ltd.

The Spaceteq is a business unit of Denel Dynamics, which is a division of government-owned Denel SOC Ltd. The parent company has its root in missile research, design and development that started in 1964 with the Institute for Rocket Research (IRR), which later became the National Institute of Defence Research (NIDR), under the CSIR. This cumulated to the establishment of Kentron (Pty) Ltd as a division of Armscor in 1978 with the aim of providing defence system and support for the military. When the Denel (Pty) Ltd was established as a group of company in 1992 and commissioned for arms exportation, Kentron became a division of this industrial group. In 2004 and 2006 restructuring of Denel, Kentron was renamed Denel Aerospace Systems (DAS) and Denel Dynamics respectively. As a unit of Denel Dynamics, Spaceteq became operational on July 1, 2013 (Denel Dynamics 2020).

Spaceteq's business interests broadly cover tactical missiles, precision-guided weapons, Unmanned Aerial Vehicles (UAVs), satellite, and integrated system solutions (IAF 2020a). Its most important space products and services include design and development of satellite systems and (both soft and hardware) subsystems or components, missions control systems, AIT assistance and testing services, as well as professional advice on space systems design and training (Spaceteq 2020). Spaceteq is located in Grabouw. It is a member of IAF since 2013. The entity equally operates in partnership with the NewSpace Systems, CSIR, DTI, DST, SANSA, UCT, University of Stellenbosch, Astroafrica Technologies, F'SATI, and Space Advisory Council (Spaceteq 2020).

Sentech SOC Ltd. is another prominent state-owned enterprise (SOE) that is involved in space-related activities in South Africa. The origin of this company can be traced to the technical division that was tasked with signal distribution of the state-owned South African Broadcasting Corporation (SABC). Sentech was incorporated as a subsidiary of SABC in 1992 and was converted into a separate public enterprise in 1996. The Sentech Act of 1966 tasked the company to provide broadcasting signal distribution services, that is, Electronic Communications Network Services (ECNS) and Electronic Communications Services (ECS), to licensed television and radio broadcasters in accordance with the South African Electronic Communications Act (Sentech 2020). The company offers television and radio broadcasting services with both terrestrial and space-based infrastructures and assets. Sentech maintain a physical presence at Honeydew in Roodepoort, Gauteng.

7.2.1.3 Development of State Institutions for Defence Space Programmes in South Africa

As earlier discussed in the last chapter, the government of South Africa has developed space-related military interests and programmes since the 1960s. However, these interests and attendant programmes were suspended between late 1980s and early 1990s, as the national security threats perception of the country changed with the political transition of the period. However, the first decade of the twenty-first century witnessed renewed interests and programmes for military space capabilities in the country. Unlike Nigeria, where space-related military interests and programme have been institutionalised with formation of DSA, South Africa does not have a specific institution at the level of agency in the DOD or service of the armed forces that are publicly known to be dedicated to operationalising its defence aspect of space policy. Nevertheless, this does not mean that there is no institution for this purpose.

Although there are limited public information on the defence space programme of the RSA, it remains the area of responsibility of the DOD. The DOD is responsible for defence policy planning and provides the strategic framework for its implementation by its agencies and different services of SADF as well as collaborative arrangement with other public and private entities. In this case, ARMSCOR is an important agency in the DOD. The ARMSCOR and its affiliated entities and partners are critical in research, design, development, acquisition, and adaption of defence technologies, including space related ones, and relevant training for personnel (Armscor 2020). Denel emerged as a division of Armscor to become a component of the separate state-owned company established in 1992. However, the agency maintains the Defence Decision Support Institute (DDSI), the Institute for Maritime Technology (IMT) and Alkantpan Test Rang among other facilities, of which some of them will be examined later in this chapter.

Different services of the South African Defence Force (SANDF), namely the South African Air Force (SAAF), the South African Army (SAA) and the South African Navy (SAN) are key to operationalise defence space assets and facilities in the RSA. Amidst these, the SAAF maintains a Joint Air Reconnaissance Intelligence Centre (JARIC), which operates ground radars with air and space ISR assets. These among other things led some experts to assume that the SAAF plays more important role in operationalising the South African defence space policy and programme. Nevertheless, there are indications that the Defence Intelligence division of the SANDF (SANDF-ID), which is also known as the military intelligence played a central role in space-related C4ISR of the RSA. Other relevant divisions of SANDF in defence space policy and programme of the country are Policy and Planning, and the Joint Operations Division (JOD).

Beyond SANDF, there are indications that some other institutions are involved in the defence and security aspects of the South African space policy and programme. These include the State Security Agency (SSA) of South Africa, which was established in 2009, when the National Intelligence Agency (NIA), the South African Secret Service (SASS), the South African National Academy of Intelligence, National Communications Centre and the Electronic Communications Security (Pty) Ltd (COMSEC) were brought together under one big structure. The SSA is responsible for all civil-related or non-military intelligence of the Republic. Within this structure, SASS is a division, which continue to be responsible for foreign intelligence and counterintelligence. In addition, the South African Police and Customs tend to be relevant in defence space policy and programme of the country.

7.2.2 Development of NGOs and Private Sector in South Africa's Space Policy

NGOs and private sector are important in the consideration of institutional development for space programmes. They are important for political and economic liberalisation of space and associated benefits. Equally, they are essential platforms for wider education or popularisation, mobilisation, and participation in space policy processes. At the same time, their expansive services make space, its technologies, and associated benefits more accessible and, in competitive environment, more affordable to greater majority of the world population. Accordingly, the development of such important institutions, platforms or sector cannot be ignored in the assessment of South Africa's space policy, its operationalisation and their contributions to the development and security of the country. Table 7.2 offers a

Table 7.2 Space-related NGOs and private companies in South Africa

Major non-state space actors	Major organisations
NGOs	ASSA, SGAC, FSDSA, SASA
Private corporations	Aerobotics, Amaya Space, Astrofica Technologies, ATE, Brolaz Projects, Cobham, Carl Zeiss Optronics, Convergence Partners, Cubespace, DeltaV Aerospace, Dragonfly Aerospace, EMSS Antennas, ETSE Electronics, Geosmart Space, Marcom, Marnic Precision Engineering, Mzansisat, Newspace Systems, Neotel, Pinkmatter Solutions, Reutech, SCS Space, Simera Sense Pty, SAC, Stone Three, Stratosat Datacom, SpesCom, Spectrem Air, SunSpace, Tellumat & Telkom

Source Compiled from The Tauri Group (2011); Space in Africa (2019a)

summary of known space-related NGOs and private companies in South Africa. This section is divided into two subsections: one on NGOs and the other on private actors in South Africa's space policy.

7.2.2.1 Development of NGOs in South Africa's Space Policy Processes

Several NGOs have emerged in South Africa with space policy as well as R&D as their explicit or implicit interests. Prominent among these organisations is the Astronomical Society of Southern Africa (ASSA). The society was born out of the merger of the Cape Astronomical Association (CAA) and the Johannesburg Astronomical Association (JAA) to form a national association known as the Astronomical Society of South Africa in 1922. The CAA was established in 1912, while the JAA was established in 1918. The Society adopted its current name, with South Africa changed to Southern Africa, in 1956. The objective of the Society is to stimulate and encourage the study of astronomy, and astronomical observation and research, as well as disseminate relevant information to the community of astronomer. The Society is an umbrella association for astronomical centres and societies in the country. The major centres of the Society are located in Bloemfontein, Cape Town, Durban, Hermanus, Garden Route, Johannesburg, and Pretoria (ASSA 2020).

The national branch of the SGAC is another major NGO with space focus in South Africa. As noted earlier, the Council is representing the interests of university students and young space professionals between ages 18 and 35 in engagements with the UN, space agencies, industry, and academic community (SGAC 2020a, b). Established in 1999, the organisation has HQ in Vienna and branches in over 150 different countries, including South Africa. Notably, the country hosted the 10th annual conference of the organisation, held in Cape Town, in 2011, which was attended by 130 delegates from 42 countries (SGAC 2020b). The national branch of this organisation maintains close relations with space-related government institutions and other NGOs in the country.

The Foundation for Space Development, South Africa (FSDSA) is another major NGO with space focus in the country. It was established in 2009 with a base in Cape Town. The objective of the Foundation is to drive space awareness, outreach, education and activities (FSD 2020a, b). The South African Space Association (SASA) is also an NGO that focus on the subject matter. This association is a forum for the professionals to exchange ideas. It maintains physical presence in Cape Town. The SASA has collaborated and partnered with local and international players in space. The association is recognised by the IAF (2020b) to maintains a membership of this INGO since 2012.

There are many other NGOs with varying degrees of interests in space and attendant technologies in South Africa. Prominent among them are the South African Society for Atmospheric Sciences (SASAS), the Wildlife and Environment Society of South Africa (WESSA), the Centre for Environmental Rights, the World Wide Fund for Nature (WWF) South Africa, and the Wildlands Conservation Trust (Wildtrust). These among other NGOs are primarily concerned with the sustainability of the environment, and to varying degrees, how they can use space to advance their core interests.

7.2.2.2 Development of Private Sector in South Africa's Space Industry

South Africa has the highest concentration of private companies in its space industry than any other country in the continent. The operationalisation of space policy and programmes of the country is considerably supported by the existence of vibrant and thriving private sector. No fewer than 23 space-related companies in South Africa were identified in a report by The Tauri Group (2011). These include two state-owned companies (that were discussed earlier in this chapter) and 21 private companies, such as Aerosud, Ansys, Advanced Technologies and Engineering Co. (ATE), Brolaz Projects, Cobham, Carl Zeiss Optronics (Pty) Ltd, Convergence Partners, EMSS, Marcom, New Dawn Technologies (NDT), Neotel, Radio Holland South Africa, Reutech Radar Systems (RRS), Reutech Communications, Space Advisory Council (SAC), Space Commercial Services (SCS), SpesCom, Spectrem Air, SunSpace, Tellumat as well as Telkom (The Tauri Group 2011).

A recent report by Space in Africa (2019a) recognised 18 space companies in South Africa, which include several companies that are not identified by The Tauri Group. These include Aerobotics, Amaya Space (Pty) Ltd., Astrofica Technologies, Cubespace, Deltav Aerospace, EMSS Antennas, ETSE Electronics, Geosmart Space, Marnic Precision Engineering, Mzansisat, Newspace Systems, Pinkmatter Solutions, Reutech Radar Systems, SCS Space, Simera Sense Pty, SAC, Stone Three, Stratosat Datacom (Space in Africa 2019a). It is important to note, however, that there is no exhaustive list of companies that are directly or indirectly involved in the space industry in South Africa. Accordingly, the aim of this section is only to examines some of the prominent and most relevant private actors in the sector.

The Advanced Technologies & Engineering Co (Pty) Ltd. is one of space-related private companies in South Africa. It is popularly known as the ATE South Africa. The company is a key industrial subsidiary of the ATE Group of

Companies, which was established by the French Aeronautical Engineers. It was established and registered in South Africa in 1984 (ATE 2020). This company is located in Halfway House, Gauteng. It is a specialist system integrator that is involved in designing, integrating and testing of avionics systems for military aircraft; ballistic algorithms and firing control systems for weapons platforms; as well as development and production of remote sensing platforms (The Tauri Group 2011).

Aerobotics is an MNC with business presence in 18 countries, including France, Germany, Italy, Mozambique, Russia, South Africa, the USA, and Zimbabwe. In South Africa, the company maintains physical presence in Cape Town. It employs data (imageries) generated from drone and satellite to support farmers with precision agriculture, especially in the areas of plantation management. The company offers services that include farm survey, problem identification, yield forecast and management (Aerobotics 2020; Space in Africa 2019a).

Aerosud is another space-related company in South Africa. It maintains physical presence in Pretoria. The company engages in aviation engineering for civil and military purposes. It is involved in researching, designing, prototyping, developing, manufacturing, supplying, fitting, and servicing aviation products, which include aerostructure and interior components of aircraft, as well as robotics and automated solutions (Aerosud 2020; The Tauri Group 2011).

Amaya Space (Pty) Ltd. is a leading South African company that is involved in space manufacturing and commercial services with special focus on nanosatellite. Established in 2018, the company is a spin-off of over ten-years involvement of the CPUT and its affiliated F'SATI in space programme, and particularly satellite research, design, development, and production. The objective of the company is to commercialise nanosatellite systems and subsystems as well as associated services and knowledge locally and globally. Amaya Space maintains a physical presence in Cape Town. Its key partners include the CPUT, F'SATI, Cycle Space (the Scottish distribution partner), Stone Three, SANSA, DST, Etse electronics, CSIR, and the University of Stellenbosch (Amaya Space 2020a, b; Space in Africa 2019a, b, c, d, e).

Ansys Limited is another private company that is relevant and involved in South Africa's aerospace industry. It ISO 9001 and CIDB certified South African company, which maintains a physical presence in Pretoria. The company engages in design, development, production, integration, and distribution of technology-driven engineering solutions in the defence, information security, industrial, mining, rail, and telecommunications sectors. The company is a founding member of the South African Electrotechnical Export Council (SAEEC) and equally affiliated to the Aerospace, Maritime and Defence Industries (AMD) and the Fibre To The Home (FTTH) Council (Ansys 2020; The Tauri Group 2011).

Astrofica Technologies Pty Ltd was established in 2018. The name was coined from Astro and Africa. The company was born from the dreams of Jessie Ndaba and Khalid Manjoo, which were slowly conceived from 2006, when they were involved in the SumbandilaSat project of the DST. Its services revolve around satellite technology solutions, particularly in the areas of assembling, integration, and testing of satellite systems. The company maintains its head office in Cape Town and an AIT facility in Grabouw. Astrofica's partners and clients include the DST, SANSA, CPUT, Amaya Space, Newspace System, Cocalibra Engineering, Simera Sense, Cube Space, and Spaceteq (Astrofica 2020; Space in Africa 2019a, b, c, d, e).

Brolaz Projects (Pty) Ltd. is a member of Brolaz Group of Companies. It was established in South Africa in 1989 with a base in Midrand, Gauteng. The Company has developed and expanded over the years, as evident with its growing number of subsidiaries across major African countries. The Brolaz is a leading infrastructure service provider in the areas of Telecommunication, Broadcasting, and electronics. It offers tailored-made services to meet the requirement of its customers in different areas of turnkey solutions for infrastructure projects. It specializes in the design, development, production, and supply of relevant infrastructure components such as towers, masts, conventional and alternate power, cooling systems, equipment shelters, as well as camouflaged antenna and the supporting structures. Its clients and partners comprise of blue-chip companies, broadcast operators, equipment vendors, fixed and mobile telecommunication operators, government bodies, oil companies as well as many other investors locally and internationally (Brolaz South Africa (Pty) Ltd. 2011; Apollo 2020; The Tauri Group 2011).

Cobham Aerospace Communication is another South African company that is involved in the aerospace sector. Although the company is known as the Omnipless Manufacturing (Pty) Limited, it is trading as the Cobham Aerospace Communications. The company maintains physical presence in Western Cape. It offers services to clients in commercial, defence and security aerospace sectors. Its operation covers a range of services, including aircraft clocks, light, satellite communications system, antennas, audio and radio management, avionics and supporting systems, as well as UAV connectivity among other tailored-made services required by customers (Cobham 2020a, b; The Tauri Group 2011).

CubeSpace is one of the key players in South Africa's space industry. It was born from the success recorded in academic research and innovation of the ESL of the University of Stellenbosch and the Innovus programme of its Nedbank LaunchLab business incubator on altitudinal determination and control systems (ADCS) in 2014. CubeSpace is officially recognised as a spin-off of the University in 2017. The company is located in Stellenbosch. Its products and services revolve around micro-satellite sub-systems, with special focus on ADCS, including onboard CubeSat computer, sensor, control, and experimental satellite cameral sensor. Its clients and partners include the NASA, the ISIS, Hawaii Space Flight Laboratory, Open Cosmos, and Alén Space (Cubespace 2020; Space in Africa 2019a).

The Convergence Partners is another company in South Africa with space-related focus. It was established in 2006. The company maintain a physical presence in Johannesburg, South Africa, and some other African countries. It provides skills, experience, and capital to accelerate technological development in the areas of communications access and ICT infrastructure development, with special focus on initiatives that raise the availability of communications and broadband services as well as new technology in Africa (Convergence Partners 2020; The Tauri Group 2011).

DeltaV Aerospace is another emerging space-related company in South Africa. It was established in 2014 and located in Cape Town. The company provides services in the areas of propulsion engineering, aerospace engineering, systems engineering, test engineering and related programme management. Accordingly, DeltaV is into research, design, simulation, development, assembling, integration, testing and supply of propulsion devices, launcher vehicle, satellite and UAV systems and subsystems (DeltaV 2020; Space in Africa 2019a).

The EMSS group of companies is another major player in South Africa's space industry. It was established in 1994 by Dr. Frans Meyer and Dr. Gronum Smith to render services in electromagnetic consulting. Over the year, the company developed a series of additional services and products, which matured into independent status that made it a group of companies. Amidst these are computational electromagnetic software that was associated with FEKO, which was sold to Altair in 2014. The EMSS Antennas was also established as one of such new entities in 2002. This entity is based in Stellenbosch. It provides antenna hardware and software and supported local and international industry with electromagnetic engineering services and solutions. Since 2006, the company has focused on developing receiver solutions for radio telescopes and offering services in the area of reflector design, which include both analysis and synthesis. When EMSS group of companies rebranded as Alphawave Group in 2017, the EMSS Antennas equally became one of its subsidiary entities (EMSS Antennas 2020; Space in Africa 2019a; The Tauri Group 2011).

The ETSE Electronics is another subsidiary of Alphawave Group (formerly EMSS group of companies). It was established in 1997 with a base in Stellenbosch. The company has developed products and offered services for space, military, agriculture and commercial applications. It is involved in research, design, development, manufacturing and consultancy in the areas of radio frequency (RF) and microwave electronic design (including at the levels of component, module, and system solutions). Its experience covers implementation of technologies such as embedded systems and software, Field Programmable Gate Array (FPGAs), Global System Mobile (GSM), GPS, RF modems, microprocessors, low-power systems, sensors, and measurement devices (ETSE 2020; Space in Africa 2019a).

GeoSmart is another South Africa's space-related company that emerged from R&D efforts of the University of Stellenbosch. It was established by Prof Adriaan van Niekerk as a spin-off company of the University's Innovus LaunchLab business incubator and is based in Stellenbosch. The company maintain close ties, partnership, and collaboration with the university as well as its affiliated entities like the Innovus and the Centre for Geographical Analysis (CGA). It employs various cutting-edge technologies, such as the geographical information systems (GIS), earth observation satellites, global navigation satellite systems (GNSS), mobile devices, cloud computing and artificial intelligence to provide geospatial solutions to geographical problems in the country (Geosmart 2020; Space in Africa 2019a).

Marcom Aeronautics and Space (Pty) Ltd. is another major player in South Africa's space industry. It maintains a physical presence in Johannesburg. Marcon's strategic partners include AISI, CSIR, SANSA and SACSA. The company is involved in research, design, simulation, development, assembling, integration, testing, manufacturing, and supply of launcher systems and subsystems. Its products and services cover a broad spectrum, including liquid rocket engine design and testing; guidance, navigation and control systems; attitude control system design; orbital mechanics, trajectory optimization; thrust vector control systems; structural dynamics; and aero-thermodynamics (Marcom 2020; The Tauri Group 2011).

Marnic Precision Engineering started as a family business founded by Pottie Potgieter in 1988. It grew from a backroom business to its facility in Johannesburg and was incorporated as a company in 2016. The company provides precision engineering products and services for aerospace and defence, medicine, agriculture, mining, power and energy, petrochemicals, and food and beverage industries. Its products and services include computer numerical control (commonly referred to as CNC) milling, turning, conventional milling, wire cutting, welding, grinding, fabrication, reverse engineering, and product development (Marnic 2020a, b; Space in Africa 2019a).

Mzansisat is another major player in South Africa's space industry. It was established in 2013 with a base in Stellenbosch. The company provides satellite broadband services and connectivity for a wide range of industries, from finance and education, to mining and agriculture, as well as the government entities, including the military, police, and disaster management agencies. It boosts of providing connectivity from a sovereign satellite, that is, the space asset that is controlled and operated within the country (Mzansisat 2020; Space in Africa 2019a).

New Dawn Technologies (NDT) is a South Africa's ICT company that was established in 1999. It maintains physical presence in Midrand. The NDT's strategic partners include Oracle, Microsoft, SAP, Vodacom, IBM, and CITRIX. The company provides systems integration and business consulting solutions across various industries, which include communication, education, energy, entertainment, government, financial, health, hospitality, information, leisure, manufacturing, mining, retail, tourism, utilities, and other sectors (NDT 2020a, b; The Tauri Group 2011).

Newspace Systems (Pty) Ltd. (NSS) is another major space company in South Africa. It was a product of a joint venture that involved the South African company, SCS Aerospace Group, and the Dutch SSBV Group. The company maintains physical presence in Cape Town, South Africa as well as in the United Kingdom. Its strategic partners and clients include AstroDigital, AstroScale, Berlin Space Technologies (BST), Aerospacelab, Conae, Hemeria, KMI, Lapan, Leo Stella, Gomspace, Nexeya, OHB, Okman Aerospace, Sierra Nevada Corporation, Sitael, Spaceflight, and Syspective. The NSS is involved in research, design, development, assembling, integration, manufacturing and supply of spacecraft components and sub-systems, and famous in the area of ADCS. The products and services offered by the company includes CubeSat attitude control solutions, closed-loop

ADCS, GPS receivers and antenna, magnetometers, magnetorquer rods, reaction wheels, sun sensors, stellar gyros, star mappers and waveguides (NSS 2020; Space in Africa 2019a).

Pinkmatter Solutions is another space-related company in South Africa. It maintains core physical presence in Pretoria, South Africa and Berlin, Germany. The company specialises on R&D of earth observation satellite, ground station and data management software. Its products and services include ground station and data management software, the FarEarth software suite, encompasses antenna, satellite downlink scheduling, high volume data processing, data archiving and cataloguing, automated services for the processing and visualization of earth observation imagery, such as high-resolution elevation models for encroachment detection, vegetation analysis as well as flood and fire monitoring. Its clients and partners include several space and remote sensing agencies or centres across Africa, Asia, Australia, Europe, and North America (Pinkmatter 2020; Space in Africa 2019a).

Reutech Pty Ltd. is a group of companies that produce and supply space-related products and services. Prominent among its subsidiaries are Reutech Radar Systems, Reutech Mining, Reutech Solutions, Reutech Communications, Reutech Electronics, Reutech Rouge International, Fuchs Electronics, Nanoteq, Omnigo, Dopptech, Terra Firma Solutions, and Bluenova Reunert Limited. As a group, these companies design, develop, manufacture, provide and supply high-precision products and services for defence, mining, industrial, and other commercial applications. Amidst these, the Stellenbosch-based Reutech Radar Systems (Pty) Ltd. (RRS), which was established in 1987 and now owned by Reunert Ltd., is involved in research, design, development, supply and installation of land and sea-based and solar radar systems and tracking devices, as well as telescope products. Reutech Solutions provides statics and mobile tactical ICT support for C4. Reutech Rouge International deals with remote control systems. Reutech Communications provides secure communication for SADF and some international clients. Other Reutech subsidiaries are equally relevant in production or provision of supporting products and services for space systems and sub-systems (Reutech 2020a, b; Space in Africa 2019a; The Tauri Group 2011).

Space Advisory Council (SAC) is one of the spin-off companies from the South Africa's satellite demonstrations: SunSat and SumbandilaSat. It was established in 2009 with a base in Stellenbosch. The company specialises in research, design, development, assembling and testing of spacecraft and payloads, with emphasise on nano and micro satellites engineering. It offers system and sub-system solutions to clients on EO satellite programming. SAC's products include optical payloads Gecko Imager, Chameleon Imager and Tegu Imager; optical satellites like phoenix (25 kg) CubeSat, Lightweight High Resolution and High Performance (1 m) Satellites and Very High Resolution (50 cm) Satellite; as well as Synthetic Apreture Rader Satellites like uStar, FalconSAR and Dragonsar. Its clients include China and Netherlands (SAC 2020; Space in Africa 2019a; The Tauri Group 2011).

Dragonfly Aerospace is a rapidly emerging new space company in South Africa. It was co-founded by Bryan Dean and Rikus Cronje in January 2020 as Southern Space and was renamed as Dragonfly Aerospace. It was established from the collapse of the SAC in late 2019. It built on space heritage developed by the Stellenbosch University, SunSpace, SAC and Denel Spaceteq. Based in the Western Cape, the company is involved in research, design, development, assembling and testing of spacecraft and payloads (Space in Africa 2020; Dragonfly Aerospace 2023).

Space Commercial Services, otherwise known as the SCS Space (Pty) Ltd is another private actor in South Africa's space industry. It was established in 2008 and maintains physical presence in Cape Town and Stellenbosch. The company provides advisory and consulting services on satellite programming. Its strategic partners are SAC and the NewSpace Systems (SCS Space 2020a, b; Space in Africa 2019a; The Tauri Group 2011).

Simera Sense Pty Ltd is another space-related company in South Africa. It was established in 2018 as a subsidiary of Simera Group. It maintains physical presence in Somerset West, Western Cape. The company specialises on end-to-end optical payload for EO satellites. It designs, develops, manufactures, and supplies different systems and subsystems or elements of a remote sensing payload, including optical front-end, control, sensor, processing power, structure, robustness and calibration. Its products include different variant of xScape100 and xScape200. The company is into partnership with Amaya Space, Clyde Space, CubeSpace, ISISpace, GOMSpace, Open Cosmos, Nano Avionics, Space Inventor, and Satsearch (Simera 2020; Space in Africa 2019a).

Spescom (Pty) Ltd is a South African ICT company that was established in 1977. It maintains physical presence in Stellenbosch, Durban, and Johannesburg. The company is made up of five divisions, including Spescom DataFusion, Spescom DataVoice, Spescom Telecommunications, Spescom Media IT and NewTelco South Africa. Spescom specialises on delivering integrated communication solutions for businesses to connect and communicate with their customers with voice, video and data technologies. It offers integrated contact centres, platforms and applications, as well as telecommunications and broadcast solutions (Spescom 2020; The Tauri Group 2011).

Stone Three is another major space company in South Africa. It was established in 2000 and based in Somerset West. The company specialises on IoT and AI-augmented solutions. It develops software-defined radio (SDR) products, maritime VHF data exchange systems (VDES) as well as automatic identification system (AIS) space receivers for microsatellites (Stone Three 2020; Space in Africa 2019a).

Stratosat datacom is another company in South Africa's space industry. It is a subsidiary of the German based Schaueburg International Group, which equally maintain subsidiaries in some other African countries like Angola, DR Congo, Mozambique, and Nigeria. Stratosat was established in 2002 with a base in Gauteng. Its interests and investment in South Africa broadly cover mining, government, transportation, enterprise, NGO, construction, defence, oil and gas, finance, health, agriculture, and ICT. The company provides turnkey satellite and microwave wireless converged communication network solutions, which entails design, production, supply, systems integration, installation, training and operational services to clients and partners, such as Internet Service Providers (ISPs), Mobile Network Operators (MNOs), broadcasters, satellite network operators and system integrators (Stratosat 2020; Space in Africa 2019a).

Other notable space related companies in South Africa are Carl Zeiss Optronics (Pty) Ltd, SunSpace, Spectra Air, Neotel, Telcom and Tellumat. The Pretoria-based Carl Zeiss focuses on rangefinders, thermal imagers, handheld targeting systems, stabilized multi-sensor platforms for use in helicopters, special aircraft and drones, and submarine periscopes. The Stellenbosch-based SunSpace produced satellite systems and sub-systems. Spectral Air specialises in air and space supported survey. Neotel and Telcom are leading communication companies in South Africa that provide broad range of services with terrestrial and space infrastructures across the country. Neotel maintains its head office in Johannesburg, while that of Telcom is based in Pretoria. Based in Cape Town, Tellumat specialises on defence, electronics, and communication technology (The Tauri Group 2011).

7.3 Development of Mobilisation Capacity for Space Policy Programmes in South Africa

As earlier noted, institutions can only give life to space policy and strategy within the scope of their mobilisation capacity. The extent of development and security objectives that are achieved with space policy and strategy is a function of relevant institutional capacity. In this case, human and material resources are critical to the development of mobilisation capacity of the institution/s that are required to develop, display and utilise space capabilities for the development and security aspirations of a country. These are essential resources for the management of organisational goals. This section is further subdivided into two subsections, which focus on the human and material resources that made up South Africa's space sector.

7.3.1 Human Resource Development in South Africa's Space Sector

Human resources are critical to operationalisation of space like any other policy and strategy as well as realisation of national and organisational objectives. These include the totality of individuals that are directly or indirectly involved in space policy implementation, R&D, and industry as well as their operational environment, skills, motivations, and performance. However, this section largely focuses and attempts a survey of personnel that are involved in space R&D, policy implementations and industry, and to a limited extent, relevant skills, in the RSA.

It is important to note that this study is unable to fully determine the aggregate number of people that are involved in this sector. Besides, previous efforts towards this direction by Space in Africa (2019a, b) are based on extensive profiling of major public institutions and private corporations that are involved in the sectors. However, the list of profiled public and private entities is not exhaustive, they excluded many entities that are indirectly involved in, and those that provide supporting services for, the sector. They equally excluded academic communities that are directly or indirectly involved in space R&D. Nevertheless, these and other efforts provide useful insight into personnel strength of different space entities in South Africa, the space sector of the country as a whole, and elsewhere across the continent. Table 7.3 offers a summary of some of the major space entities and the known total number of their employees in South Africa.

Table 7.3 Major Space entities and the total number of their employees in South Africa

No	Entity	Total employees	
		As reported by Space in Africa (2019a)	Officially reported or other sources, 2019/20
State owned			
1	SANSA	–	177
2	Denel Spaceteq	26–50	–
3	Sentech Soc Ltd	>500	200
Private corporations			
4	Aerobotics	56	48
5	Amaya Space	11–25	–
6	Astrofica Technologies	10	–
7	Cubespace	10	17
8	DeltaV Aerospace	63	–
9	EMSS Antennas	11–25	–
10	ETSE Electronics	11–25	13
11	Geosmart Space	1–10	–
12	Marnic Precision Engineering	26–50	35
13	Mzansisat	1–10	–
14	Newspace Systems	22	–
15	Pinkmatter Solutions	11–25	–
16	Reutech	51–100	–
17	SCS Space	1–10	–
18	Simera Sense Pty	15	–
19	SAC	47	–
20	Stone Three	40	–
21	Stratosat Datacom	30	–
22	Xinabox Ltd	10	–
23	Dragonfly Aerospace	–	20

Source Space in Africa (2019a) and author's extension

Unlike NASRDA that has little publicly available information on its human resources and their management, SANSA provides a detailed annual report on the subject matter. Between 2013 and 2019, SANSA maintains between 148 and 183 permanent staffs (SANSA 2014, 2015, 2016, 2017, 2018a, b, 2019a, b, c, d). As at March 2019, SANSA has 177 personnel, which fell short of approved 203 personnel for the agency, leaving it with 26 (12.8%) vacancy (SANSA 2019a). Although it represents an increase from 148 staffs in 2014, it was a decline from 183 staffs in 2018. As at March 2019, the number of SANSA's employees as categorised by salary levels are one person at the position of top management, six at senior management level, 56 professionally qualified, 93 skilled, 20 semi-skilled, and one unskilled. In term of employment by programme, the agency was manned by 37 employees for administration and space engineering, 31 staffs for Earth Observation, 55 personnel for space science and 54 employees for space operations as at 2019 (SANSA 2019a).

The SANSA equally maintains some public records that are relevant to assess its level of personnel diversity, discipline, retention, training, and morale (SANSA 2014, 2015, 2016, 2017, 2018a, 2019a). In term of gender, the personnel strength of the agency is made up of 105 (59%) male and 72 (41%) female. In term of race, SANSA maintains 112 (63.3%) African employees, 16 (9%) coloured, 11 (6.2%) Indian, and 38 (21.5%) White. Although the Indian appears to be the least in number in the agency, they can be found in the positions of top management and senior management. Moreover, 66.7% (48) of the female employees of the agency are Africans, while 20.8% (15) are Whites, 9.7% (7) are coloureds and 2.8% (2) are Indians (SANSA 2019a). These among others considerably show the level of diversity of the agency.

SANSA's records of retention are also instructive. Between 2013 and 2014, the agency terminated the appointment of 8.8% (13) of its 148 permanent employees (SANSA 2014). Between 2014 and 2015, the appointments of 13.1% (21) of its 184 personnel were also terminated. Between 2013 and 2015, temporary or contract staffs also declined from 36 to 23, as permanent staffs increase from 138 to 160. Between April 1, 2018, and March 31, 2019, there were equally 12 appointments and 19 (10.7%) termination of appointment in the agency. Amidst these, the termination of appointment is connected to nine resignations, seven expiration of contract, two retirements and one dismissal between 2018 and 2019. At the same time, the agency responded to misconduct with three written warning, three final warning and one dismissal (SANSA 2019a). These among other offers some insight into labour relations: recruitment, retention, turnover and discipline within the agency.

Personnel welfare, morale and training are equally important in agency's human resource development and management. The SANSA spent 48.6% of its total expenditure on personnel in 2018/2019. On average, R653.674 million were voted for personnel cost per person during this period. Amidst these, 0.89% of the personnel cost was voted for training. The SANSA trained or retrained 93.8% (166) of its total personnel in 2018/2019. Accordingly, R8.239 million was used in training personnel for the period (SANSA 2019a). These among others suggest the existence of a robust welfare package and training programmes, which are expected to boost personnel skills and morale for optimum performance, in SANSA.

Beyond SANSA, there are many other organisations that make up or support the space industry in implementing relevant policy and conducting related R&D. The state-owned Denel Spaceteq is reported to have between 26 and 50 employees in early 2019. Another state-owned company, Sentech was reported to have over 500 personnel by Space in Africa (2019a), although Dun & Bradstreet (2020) estimated that the company has 200 employees. Table 7.3 provides a summary of the private actors in space industry and the number of their employees as reported by Space in Africa (2019a). However, there are cases where officially given information in late 2020 differ from these data. For instance, Cubespace was estimated by Space in Africa (2019a) to have between 1 and 10 employees, while the company officially recognised 17 personnel as member of it teem (Cubespace 2020). Space in Africa (2019a) also reported that ETSE Electronics has between 11 and 25 employees. However, the company officially recognised 13 personnel as of 2020 (ETSE 2020). Equally, Space in Africa (2019a) reported that Marnic Precision Engineering has between 26 and 50 employees. Nevertheless, the company officially recognised 35 employees in 2020 (Marnic 2020a). Dragonfly Aerospace was also reported to have about 20 staffs as of late 2020 (Space in Africa 2020).

Beyond these, tens of thousands of people are associated with space and related NGOs in South Africa. These are in additions to thousands of scientists and engineers that are engaged in space R&D and related fields across South African universities as well as other high degree awarding institutions and research centres. Moreover, these systems are producing tens of thousands of scientists and engineers annually, with the potentials to contribute directly or indirectly to space policy processes (advocacy, formulation, and implementation), R&D, and industry. These are separated from the thousands of scientific enthusiasts in space and other related areas with background in art, social sciences, law, and business management in the country. These among others provide considerable insight into actual and potential human resources that are available for mobilisation of the republic in the quest for space capabilities and associated policy goals.

7.3.2 Material Resources in South Africa's Space Sector

Beyond human resources, material resources are also critical to every organisation and their performance. Accordingly, the material resources that are available to relevant institutions to operationalise space policy in South Africa cannot be ignored in this analysis. This section therefore examines the material resources that underscored space policy and industry in the RSA, with special attention for the financial resources and relevant facilities.

7.3.2.1 Space Expenditure in South Africa

Funding is an important element of material resources that determine institutional capacity, including in the areas of personnel strength, skills, and morale as well as facilities that can be devoted to their operations. Since the end of Apartheid regime, South Africa's civil space programmes have emerged out of its military shadow, which makes it more possible to assess its expenditure. Although the country retains a military space programme, of which its details remain less accessible for many research attempts, including this one, its civil programmes have become more defining in public and academic focus on its space policy.

There are multiple sources of data on South Africa's space expenditure. For instance, data from Euroconsultant offers some insights on the civil space expenditure of countries that spend more than US$ 10 million across the world, including the RSA. In this case, South Africa spent an aggregate of US$ 110 million on space in 2013, 2014, 2016 and 2018 (Oyewole 2020a). The country spent US$ 23 million in 2013, US$ 31 million in 2014, US$ 20 million in 2016, and US$ 36 million in 2018. These made the country the 6th, 5th, 7th, and 6th highest space spender in Africa in 2013, 2014, 2016 and 2018 respectively (Bochinger 2016; Euroconsultant 2017; Lal et al. 2015; Oyewole 2020a; Seminari 2019). In all these cases, the government of South Africa spent less than its counterparts in Algeria and Nigeria, and in some cases, lesser than Egypt, Congo, Morocco and Angola (Oyewole 2020a).

In 2018, a Business & Market Analysis by Space in Africa (2018) shows that South Africa voted R138 million (US$ 9.7 million) to SANSA, which was 1.7 per cent of the total budget of the DST. In addition, R709 million (US$ 50 million) was voted to SKA project. These made a total of R847 million (US$ 60 million) budgeted for civil space activities, which represented 10.9 per cent of the DST's total budget. The report observed that SANSA's 2018a, b budget was a significant increase from the R131 million (US$ 8.6 million) budget of the agency in 2017 (Space in Africa 2018). However, absence of a specialise institution for space in the defence department is recognised in the report as the challenge in tracking the funds that is specifically designed for defence space programme of the country.

The financial reports of SANSA also offer another perspective to funding that is available to space activities in the country. Figure 7.1 shows SANSA's expenditure, paying attention to the approved budget, final budget, and the actual amount on a comparable basis from 2013 to 2019. The entity like others in the country maintains a finacial year that covers April 1 to March 31. This makes it a challenging task for those reports that generate space expenditure of the country on the basis of calendar year. In all cases presented in Fig. 7.1, the approved budget for the entity is lesser than its final budget, which is also not the same with the actual amount of its expenditure on comparable basis. Unlike the case of the NASRDA that the financial reports are not available to verify the claim disparity between approved budget and allocation, available data between 2013 and 2019 capture these disparities for the SANSA. These raised some questions against reliance on approved budget in assessing government expenditure on space activities among other things in South Africa and beyond.

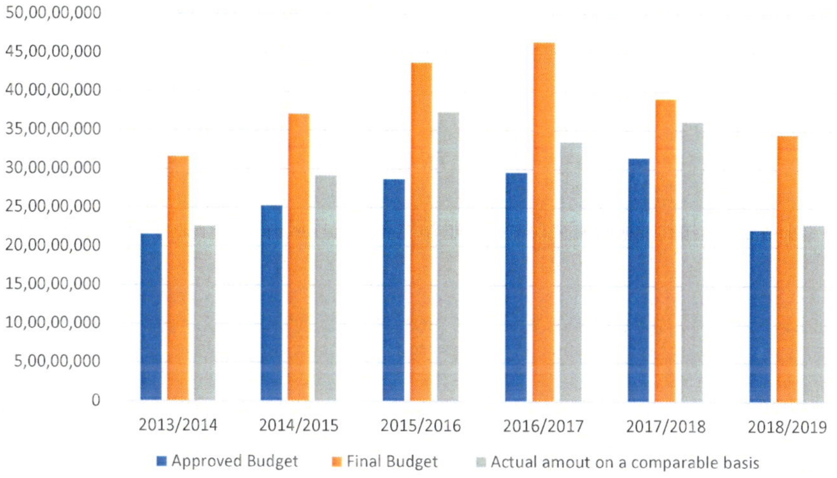

Fig. 7.1 Space Expenditures in South Africa: SANSA's budget, 2013–2019 (R). *Sources* Data extracted from SANSA (2014, 2015, 2016–2019)

Available data on SANSA's expenditure as evident in Fig. 7.1 shows that there was an increase in approved budget from R 215.6 million in 2013/2104 to its peak of R 315 million in 2017/2018, before declining to R 221.56 million in 2018/2019. The final budget for the entity rose from R 316 million in 2013/2014 to its peak of R 464 million in 2016/2017, before declining to R390.8 million in 2017/2018 and R 344.8 million in 2018/2019. Actual amount on a comparable basis rose from R 225.876 in 2013/2014 to its peak of R 373 million in 2014/2015, it declined to R 335 million in 2015/2016, rose again in R 360.957 million in 2017/2018, and declined again to R 228.76 million in 2018/2019 (see Fig. 7.1).

The report of SANSA (2019a, b, c, d) offers some unique perspective to the funding of the entity, as it paid some attention to its expenditure based on programmes. This detail was captured under human resource management and not financial or audit report. This detail, however, introduce some confusion, as the expenditure of the entity was reported in billions and not millions. The total expenditure of the entity between April 1, 2018, and March 31, 2019 was reported to be R 238.5 billion, of which R 115.7 billion (48.6%) are personnel expenditure. Amidst these, 32% of the total expenditure was voted to Space Operations, while Earth Observation, Administration and Space Engineering, and Space Science received 28%, 22% and 18% respectively (see Fig. 7.2).

Beyond SANSA, there are many other public institutions, of which funding them have direct or indirect contributions to space in South Africa. These broadly cover civil, military commercial entities owned by the government. Financial and audit reports of the earlier mentioned and other relevant state-owned institutions that are directly or indirectly involved in space activities in South Africa may offer us a fresh perspective into space expenditure of the country. This called for a revaluation of the prevailing model or metric of assessing space expenditure

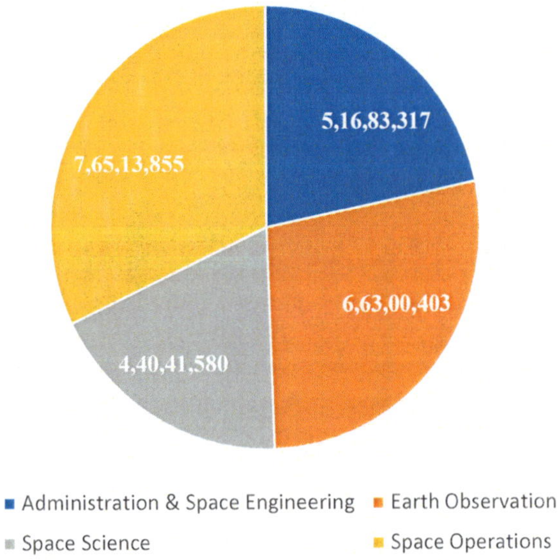

Fig. 7.2 SANSA's expenditures by programmes in 2018/2019 (R'000). *Source* Data extracted from SANSA (2019a, b, c, d)

of the country. In doing this, research efforts can be channelled to determine the aggregate or relevant aspect of disaggregate expenditure of concerned entities.

Assessment of space expenditure in South Africa will not be completed without special attention for private capital in the industry. Private companies that are involved in space activities in the country has generated funds for their operations through multiple sources, including personal saving/earning of the founders, grant, loan, investment, and stock market. For instance, equity investment in South Africa's space industry in the last decade include the US$ 4.6 million raised by the Aerobotics, the US$ 2 million raised by Simera Sense Pty Ltd, and the US$ 500,000 raised by GeoSmart (Space in Africa 2019a). These among others are indicators of private investment and expenditure in space activities in the RSA.

7.3.2.2 Ground Facilities and Infrastructural Development for Space R&D in South Africa

Ground facilities are important aspect of space and other scientific R&D capabilities as well as general infrastructural development for national growth, development, and security. The facilities are products of funding or financial investment committed to the realisation of space policy. However, facilities like land, building or office spaces and relevant equipment are critical to derive optimum utility from human resources in operationalising the policy in question. Accordingly, this subsection examines some of the available facilities, especially the location of office spaces and relevant equipment installed, for space R&D in South Africa.

South Africa is home to several facilities for space R&D, which are disperse across the country. These include facilities owned by state institutions and private corporations as well as those jointly held with international partners or consortium. As the primary space actor in the country, SANSA maintains three major facilities. One of them is located at Enterprise Building, Mark Shuttleworth Street, Innovation Hub, Pretoria 0087, Gauteng, South Africa (see Fig. 7.3). This is where the agency sites its administrative head office as well as Space Observation and Space Engineering Programmes. The SANSA has another facility for its Space Operations Programme, which is located at Farm No 502 JQ Hartebeesthoek, District Krugersdorp (see Figs. 7.4, 7.5 and 7.6). The third facility is maintained for Space Science Programme at Hospital Street, Hermanus 7200, Western Cape, South Africa (see Fig. 7.7). Figures 7.3, 7.4, 7.5 and 7.7 present selected satellite imagery of the locations and structures or buildings that make up these facilities.

For the purpose of Space Operation, Hartebeesthoek facility is very important for SANSA. The facility is equipped with antenna systems that are sited across 40 locations on the 4000 ha of land. These include L, S, C, Ext C, X, Ku, DBS, and Ka frequency bands. Amidst these are S-band for mobile support, and several other antennas for full-motion TT&C. These facilities have the capabilities of providing antennas services, launch support services, TT&C services, antennas hosting and maintenance services, as well as teleport services (SANSA 2020b). For the purpose of Space Science, SANSA maintains facilities and equipment like the Space Weather Centre (SWC), Magnetic Clean Facility (MCF), Helmholtz Coil, Temperature Chamber, Non-magnetic Screening Chamber, Absolute House, Superconducting Quantum Interference Device (SQUID), as well as Science Center and Mobile Space Lab (SANSA 2020b).

Several astronomical facilities are maintained across South Africa by public, private and/or international institutions. Amidst these, Bloemfontein, Cape Town, and Sutherland host majority of South Africa's telescopes. Bloemfontein hosts Boyden observatory with 0.41 m Nishimura reflector, Watcher Robotic reflector of 0.41 m, Alvin Clark Refractor of 0.33 m, Metcalf Photographic Triplet Refractor of 0.25 m, Coelostat Telescope of 0.20 m. The likes of Elizabeth Telescope with 1.0 m; McClean Refractor with 0.61 m; Dall-Kirkham Reflector with 0.45 m; Parks Telescope of 0.41 m are based in Cape Town. In Sutherland, there are SALT, which is 11.1 m (m) by 9.8 m; Radcliffe Observatory, which is a 1.9 m telescope; Infrared Survey Facility (IRSF), SuperWasp, BiSON Telescope, Kilodegree Extremely Little Telescope (KELT), 1.4; 0.75 and 0.5 m Telescopes (AfAS 2020; Dubow 2019; Froehlich and Siebrits 2019; Oyewole 2017, 2020a). These are in addition to VLBI, geodesy, and radio telescope in Hartebeeshoek. The construction of SKA, of which buck of its facility is schedule to be located in the RSA, is another major addition to astronomy in the country.

South Africa equally have several manufacturing and AIT facilities for spacecraft systems, subsystems, and sub-subsystems. Besides AIT of satellite systems, most satellite subsystems or components are manufactured in the country. For instance, Cubesapce maintains an AIT facility for satellite subsystem and components like ADCS in Stellenbosch, while the NNS maintains a facility that serve similar purposes in Cape Town (Cubespace 2020; NNS 2020). Marnic Precision Engineering equally maintains 2600m2 manufacturing plant in Johannesburg, which provides the platform to deliver quality precision parts for different products, including spacecrafts (Marnic 2020a). More importantly, Astrofica and the

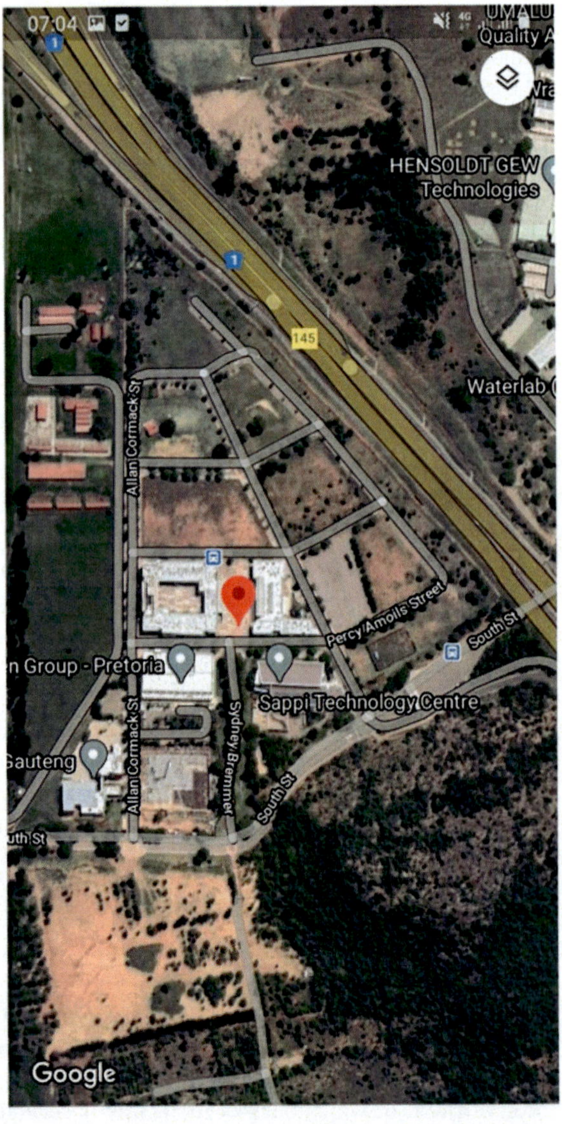

Fig. 7.3 Satellite imagery of SANSA's facility in Pretoria. *Source* Tracked and captured by the author using GPS, Google Map (2020)

SAC maintains AIT facilities for satellite systems in Graouw and Stellenbosch respectively (SAC 2020; Astrofica 2020). DeltaV Aerospace Ltd also has AIT facility in Somerset West. Dragonfly Aerospace is equally investing in a 3000 m^2 design and manufacturing facility with 1000 m^2 of cleanroom areas for microsatellite constellation production, with a plan to produce 16–48 satellites per year (Dragonfly Aerospace 2023). These are connected or in addition to the AIT and

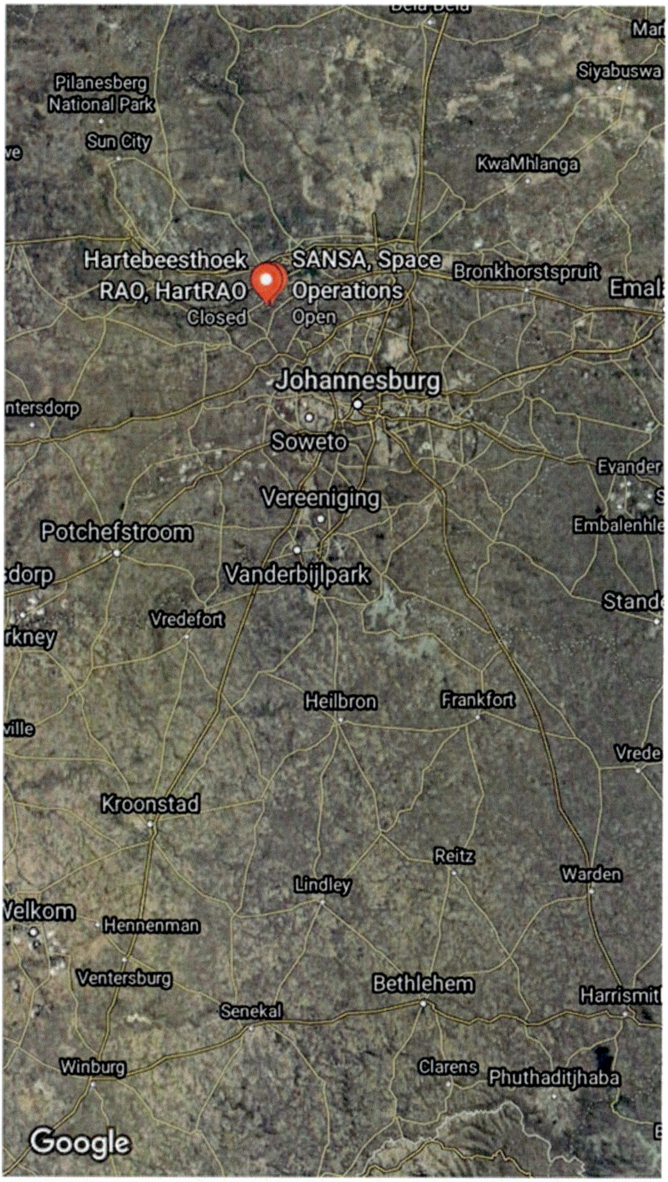

Fig. 7.4 Satellite imagery of the locations of SANSA's Space Operations and HartRAO, Hartebeesthoek. *Source* Tracked and captured by the author using GPS, Google Map (2020)

Fig. 7.5 Satellite imagery of SANSA's Space Operations, Hartebeesthoek. *Source* Tracked and captured by the author using GPS, Google Map (2020)

related facilities maintained for R&D capacities in satellite systems and subsystems by the University of Stellenbosch, UCT and CPUT. Besides, some of these universities equally own and operate satellite ground stations.

Many facilities for rocket, missile and space launcher research facilities can also be found in South Africa. Somerset West-based DeltaV Aerospace owns an 800 m^2 test facility and a 500 m^2 laboratory for research, design, simulation,

Fig. 7.6 Satellite imagery of HartRAO, Hartebeesthoek. *Source* Tracked and captured by the author using GPS, Google Map (2020)

development, assembling, integration, and testing of aerospace systems and subsystems, including launcher vehicle, satellite, and UAV (Space in Africa 2019a). Marcom Aeronautics and Space (Pty) Ltd also maintains facility for research, design, simulation, development, assembling, integration, testing, and manufacturing of launcher systems and subsystems in Johannesburg (Marcom 2020). The Faculty of Engineering of the UKZN and its Aerospace Systems Research Group (ASReG) equally own and maintain facility for Phoenix Hybrid Sounding Rocket Program (HSRP). Besides, Denel owns and maintains (orbital) launch pads and tracking systems at its Overberg Test Range near Arniston, Western Cape. Armscor equally maintains Alkantpan Test Range, which is a strategic ISO 9001 certified all-purpose ballistic test range for medium and large calibre of weapons and munition (Armscor 2020). These among other facilities are important to the operationalisation of South Africa's space policy and the realisation of its objectives.

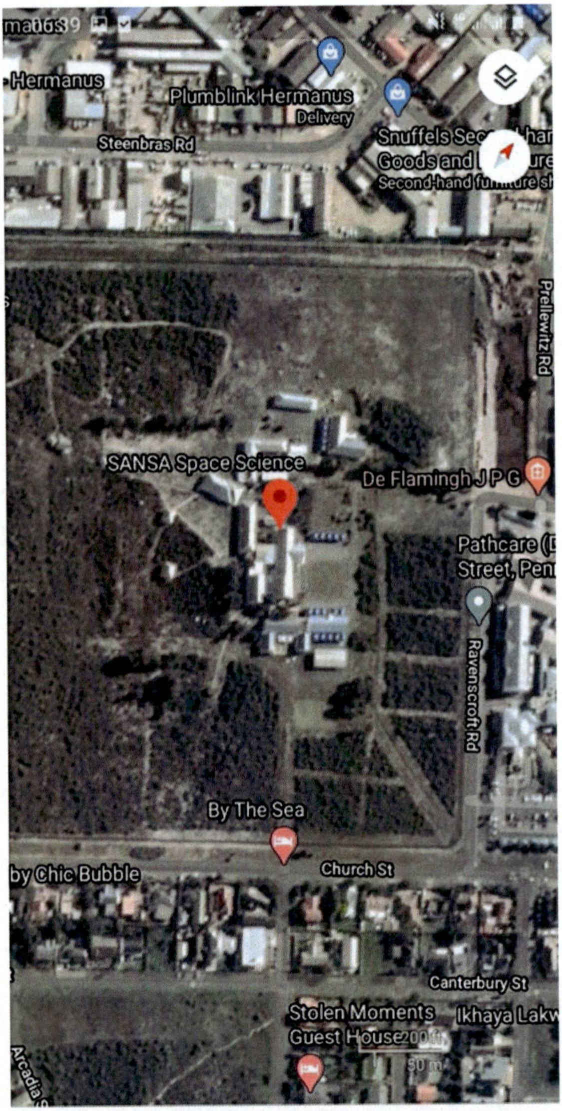

Fig. 7.7 Satellite imagery of SANSA Space Science, Hermanus 7200, Western Cape. *Source* Tracked and captured by the author using GPS, Google Map (2020)

7.4 Development of Space Access and Assets in South Africa: Satellites and Launch Capabilities

This section focuses on the level of space access and assets as elements of capabilities or resources that are available to South Africa to operationalise its space policy and realise associated development and security objectives. Accordingly, this section is sub-divided into two subsections: the first one deals with the development of space-based capabilities in South Africa, while the second one focus on the nation's space launch capabilities.

7.4.1 Development of Space-Based Capabilities in South Africa

South Africa has developed significant space-based capabilities, which reflect its regional hegemonic ambition, power and status by African standard (Froehlich and Siebrits 2019; Ogunnubi and Oyewole 2020). The country has sponsored or co-sponsored the orbiting of no fewer than twelve satellites as at the middle of 2023 (see Table 7.4). Hence, South Africa accounted for approximately 21% of the 57 satellites that were sponsored or co-sponsored from the continent as at the middle of 2023. With thirteen satellites orbited to its credit at the middle of 2023, Egypt is the only country in the continent that have sponsored a greater number of satellites into orbit than South Africa (Oyewole 2023a). At the same time, Algeria and Nigeria are behind South Africa in terms of number, with six and seven satellites respectively. With its first satellite launched in 1999, South Africa became the second African country to sponsor, own and operate a space-based asset, following the record of Egypt in 1998 (Oyewole 2017, 2020a; Froehlich and Siebrits 2019; Space in Africa 2019d). Table 7.4 provides a summary of satellite capabilities sponsored, owned, and operated by South Africa and their details.

As evident in Table 7.4, South Africa has sponsored satellites of different forms and size for different purposes. In consideration of size, the country is known to have sponsored mini-satellite (100–500 kg) such as AgriSat-1; microsatellites (10–100 kg) such as SunSat and Sumbandila; nanosatellites (1–10 kg) such as Za-Cube-1&2, Nsight-1, Za-AeroSat and MDASat-1A, B &C; as well as a picosatellite (0.1–1) such as XinaBox ThinSat. Although almost all satellites sponsored by South Africa are designed for EO, they have notable similarity and differences in their purposes. With exception of Condor-E2 and to a limited extent Nsight-1, other South African satellites are designed for educational purposes at different levels and to demonstrate the technological capacity of the country. It is not surprising that local academic and research institutions and spinoff companies are central to the research, design, development, assembling, integration, and testing of these satellites. However, South Africa unlike Nigeria, Egypt and to an extent Angola is yet to sponsor, own or operate a state-owned communication satellite in orbit.

Unlike other space nations in Africa, South Africa has displayed capacity to produce its satellites locally at least on eleven different occasions between 1990s and 2023. Besides Condor-E2 that was produced by the Russian NPO Mashinostroyeniya, every other satellites sponsored by South Africa were designed and manufactured locally. University of Stellenbosch with its Sun Space and the CPUT with F'SATI were directly involved in 66.7 per cent of all satellites acknowledged to have been sponsored by South Africa. SunSat was produced by the University of Stellenbosch, while Sumbandila and Za-AeroSat were produced by spin-off companies of the University, that is, SunSpace and CubeSpace respectively. The

Table 7.4 Development of Nigeria's Satellite Inventories and Capabilities

Satellite names	Mass of satellites (kg)	Manufacturer	Purposes	Launch & operational dates	Launch facility & locations	Cost (US$ million)
SunSat	64	University of Stellenbosch	EO/ Education	23/2/ 1999–19/ 1/2001	VAFB, California, USA	2
Sumbandila	82	SunSpace & University of Stellenbosch	EO/ Technology demonstration	23/9/ 2009–8/ 2012	Baikonur Cosmodrome	3
ZaCube-1 (TshepisoSAT)	1.2	CPUT	EO/ Technology Development/ Educational	21/11/ 2013-	DAB, Russia	–
Condor-E2	–	NPO Mashinostroyeniya	EO/ Military	19/12/ 2014-	Baikonur Cosmodrome	108
Nsight-1	3	SCS Space	EO/ Commercial	26/5/2017-	ISS	–
Za-AeroSat	4	Stellenbosch University	EO/ Educational	18/5/2017-	ISS	–
Za-Cube-2	4	CPUT	EO/ Educational	27/11/ 2018-	Vostochny Cosmodrome	–
XinaBox ThinSat	<1	High Schools	Educational	17/4/ 2019–4/ 2019	Mid-Atlantic Regional Spaceport, Wallops	–
MDASat-1A (Maritime Domain Awareness)	2/3	CPUT	EO/ Maritime Domain Awareness	13/1/2022	Cape Canaveral SFS, Florida, USA	1.7
MDASat-1B	2/3	CPUT	EO/ Maritime Domain Awareness	13/1/2022	Cape Canaveral SFS, Florida, USA	
MDASat-1C	2/3	CPUT	EO/ Maritime Domain Awareness	13/1/2022	Cape Canaveral SFS, Florida, USA	
EOS-SAT 1 (AgriSat 1)	178	Dragonfly Aerospace	EO/ Agricultural monitoring and analysis	3/1/2023	Cape Canaveral SFS, Florida, USA	–

Source Compiled by the author with data from UCS (2020); Oyewole (2017); Gunter's Space Page (2020), Space in Africa database, and author's extension

CPUT-based F'SATI equally produced the Za-Cube 1&2 and MDASat-1A, B&C. The UCT and the NMMU have equally contributed to the success of some of these and other satellite projects in South Africa. The involvement of 16 High Schools from Western Cape in XinaBox ThinSat have driven space R&D and STEM generally further to the next generation (Space in Africa 2019c; Wener 2018).

Spin-off companies from satellite R&D in universities have further expanded the coast of commercial space industry in South Africa. In this manner, the South Africa's Nsight-1 is considered as the first satellite sponsored, owned, and operated by African-based commercial entity. The satellite is a joint investment of the SCS Aerospace Group and Pinkmatter Solutions. Some of the outsourced components of the satellite like the ADCS was produced by the CubeSpace, a spin-off of University of Stellenbosch, while NMMU developed the Radiation Mitigation VHDL Coding Technique. The satellite system was assembled in the NewSpace System's cleanroom as well as integrated and tested in SAC's AIT facility. These two companies are parts of SCS Aerospace Group of Companies (Space in Africa 2019d).

The local content of most South African sponsored satellites, among other things, is largely responsible for the cheap cost of producing them, when compared to those operated by Nigeria and other African countries, without accounting for the technical variations (Oyewole 2017). Notably, SunSat cost US$ 2 million, while Sumbadila cost US$ 3 million. It was also reported that the nanosatellite programme of CPUT, which produced Za-Cube 1&2, cost the DST R16.5 million (Space in Africa 2019e). At the surface level, these are cheaper than the Russian made Condor-E2, without accounting for the technological variations in them. However, it is important to also note that most the satellites sponsored and manufactured in South Africa lasted in orbit for few weeks/months and at most few years, which is typical of micro, nano and pico- satellites designed for educational and technological demonstration purposes. However, these are shorter in lifespan when compared to 10 to 15 years' serving inventory of microsatellites purchased by Algeria, Egypt, and Nigeria. This suggest that the South Africa requires constant replacement for these satellites in a drive for self-reliance in satellite data and their utilisation, especially in comparison to other leading African nations. However, the rise of Dragonfly Aerospace, the size of its AgriSat-1 and its planned production line that is schedule to produce satellites of 100 kg to 300 kg are changing this dynamic (Dragonfly Aerospace 2023).

It is also remarkable that the first African satellite for military use, that is Condor-E2, was sponsored by South Africa. This satellite was manufactured by Russian and launched in 2014. It was a product of a US$100 million contract negotiated by the South African government with its Russian counterpart and signed on May 19, 2006 (Oyewole 2020a). The spy satellite was largely conceived, designed, and executed in secrecy, and was codenamed "Project Flute" and "Consolidated Flute", until January 2014 when it was uncovered by David Maynier, a member of the opposition party (Democratic Alliance) in the national parliament. The satellite, however, reflects South Africa's regional power and its changing defence and security realities (Oyewole 2020b, 2023b, c).

Beyond EO satellites, South Africa is yet to sponsor or co-sponsor any other space-based capabilities, most especially interplanetary mission or space station. However, South African based ground facilities or space infrastructures, most especially TT&C facilities have supported many interplanetary or extra-terrestrial missions. A national of the country had equally visited the ISS. Mark Shuttleworth, with dual citizenship of Britain and South Africa, became the first African citizen and the second space tourist to be launched to orbit and occupied ISS in 2002. The privately funded space tourism cost Mr Shuttleworth about US$ 20 million (Oyewole 2017). Aside from this, there is little or no indication by the South African government to sponsor astronauts or any other scientists or tourists to space.

7.4.2 Development of Space Access Capabilities in South Africa: Space Launch Vehicles

South Africa currently lack capability for space access, that is, to launch a satellite into the orbit. As evident in the previous Table 7.4, all the satellites sponsored from South Africa, like other African countries, were launched into orbit for the country from elsewhere. South Africa relied on Russia and the USA for orbital launch services. SunSat was launched on board a Boeing Delta II launch vehicle from the US Vandenberg Airforce Base (VAFB) in California. Sumbandila and Condor-E2 were launched from Russian leased and operated Baikonur Cosmodrome in Kazakhstan, while Za-Cube-1 & 2 were respectively launched from DAB and Vostochny Cosmodrome in Russia. Nsight-1 and Za-AeroSat, were released into orbit from the ISS, after they were launched aboard Atlas V from the USA. The Virginia Space, which owns and operates the Mid-Atlantic Regional Spaceport that is located on Wallops Island, was also contracted to provide launch service for the XinaBox ThinSat. Equally, MDASat-1A, B&C and AgriSat-1 were lunched on-board of Falcon 9 by SpaceX from the Cape Canaveral Space Force Station, USA. To have idea of the price, Sumbandila was launched at the of cost R8.3 million (Martin 2012).

Despite the foregoing concern, South Africa has committed some efforts to develop its launcher capacity. As earlier mentioned, the country has a rich space tradition, including in the area of medium-range missile and launcher system development. However, the end of apartheid regime witnessed a significant review of this programme, which is now limited to the production of tactical missiles and maintenance of facility for orbital launcher like Overberg Test Range. In the last two decades, however, South Africa has revived its interest in rocketry and launcher systems. Unlike the initial rocketry and space launcher programme that was motivated, controlled, and managed by the military for security purposes, the recent one is motivated, controlled and managed by civilians for R&D purposes.

The interests and investment are driven by the DST, SANSA, academic/research institutions, and private organisations.

Prominent among the academic institutions that are involved in rocketry and space launcher programme is UKZN. The university based ASReG, which is connected to Faculty of Engineering is equipped with laboratory-scale test facility and Mobile Rocket Launch Platform (MRLP). The group has tested three sounding rockets, including Phoenix 1A, Phoenix-1B, and Phoenix 1B II as of 2020. The last launch of the group reached 24.8 km range (ASReG 2020). Prominent among the private organisations that are involved in rocket R&D in South Africa is Marcom. The organisation has invested on CHEETAH-1 CSLV, a two stage, liquid propelled, expendable launch vehicle capable of delivering small and medium sized payloads to low Earth inclined, polar and sun-synchronous orbits (Marcom 2020). These among other organisations and activities, couple with existing infrastructures, may redefined the future of rocket R&D and orbital launch prospects of South Africa.

7.5 Conclusion

This chapter examines various institutions, including public and private entities, that are key to South Africa's space policy processes, the resources that are available to them, including the funds, human, skills and materials (facilities and equipment), and their capacities to mobilise and covert these into space capabilities (assets and access) of the country. Beyond different government entities for civil, commercial and military space programmes and NGOs, South Africa has the most deregulated space industry with flourishing private sector in the continent. Building on its space heritage, the government supported science and technological innovations in academic institutions have generated spin-off private companies that are playing critical roles in the realisation of the targeted objectives in the national space policy and strategy.

With minimum state, South Africa has been able to maximise the market prospects of space activities. This is evident in the discussion on the capabilities of the country in space assets and access, which has witnessed significant involvement of the academic community and private sector. Accordingly, this discussion on space capacities and capabilities of South Africa is key to bridge the last chapter, which assessed the policy and strategic interests of the country in space, and the next chapter that examines the extent that such objectives have been realised.

References

Academy of Science. 2020. About. Available at https://www.assaf.org.za/index.php/about-assaf.

Aerobotics. 2020. Intelligent tools for perennial crops. Available at https://www.aerobotics.com/?identifier=default-sign-up-button.

Aerosud. 2020. Aerosud. Available at https://www.aerosud.co.za/home.

Brolaz South Africa (Pty) Ltd. 2011. Brolaz towering ahead. Engineering News.
African Astronomical Society (AfAS). 2020. African Observatories. https://africanastronomicalsociety.org.
Ansys. 2020. About Us. Available at http://v2.itweb.co.za/company/ansys-ltd/?page=about.
Apollo. 2020. Brolaz SA (Pty) Ltd. Available at https://www.apollo.io/companies/Brolaz-SA-Pty-Ltd/55f22f21f3e5bb0b7c0025a1?chart=count
Armscor. 2020. What We do. http://www.armscor.co.za.
ASSA. 2020. About the Astronomical Society of Southern Africa. Available at https://assa.saao.ac.za/about/.
ASReG. 2020. About. https://aerospace.ukzn.ac.za/about/.
Astrofica. 2020. Home. Available at https://www.astrofica.com/.
ATE. 2020. Advanced technologies and engineering. Available at https://www.linkedin.com/company/advanced-technologies-and-engineering/about/.
Bochinger, S. 2016. Space exploration in the space economy. Paper presented at Symposium on the Role of Industry in Space Exploration, United Nations, Vienna, February 15.
Cobham. 2020a. Our company. Available at: https://www.cobhamaerospacecommunications.com/our-company/who-we-are/.
Cobham. 2020b. Capabilities. Available at: https://www.cobhamaerospacecommunications.com/our-company/capabilities/.
Convergence Partners. 2020. Who we are. Available at https://www.convergencepartners.com/about.
CSIR. 2020. The CSIR in brief. Available at https://www.csir.co.za/csir-brief.
Cubespace. 2020. Overview. Available at https://www.cubespace.co.za/about/.
DeltaV. 2020. https://www.deltav-aerospace.com/.
Denel Dynamics. 2020. Company profile. Available at http://www.deneldynamics.co.za/about-us/company-profile.
DHET. 2020. About us. Available at https://www.dhet.gov.za/SitePages/AboutUSNew.aspx.
Dragonfly Aerospace. 2023. About. https://dragonflyaerospace.com/about/. Accessed on Sept 15.
Dubow, S. 2019. 200 years of astronomy in South Africa: From the royal observatory to the 'Big Bang' of the square kilometre array. *Journal of Southern African Studies* 45 (4): 663–687.
Dun & Bradstreet. 2020. Sentech. Available at https://www.dnb.com/business-directory/company-profiles.sentech.ea8c89245d2a2b05699534ea7b191cc2.html.
EMSS Antennas. 2020. History. Available at https://www.emssantennas.com/.
ETSE. 2020. Welcome to ETSE. Available at https://etse.co.za/#about.
Euroconsultant. 2017. Government spending in space programs reaches $62 Billion in 2016, May 30.
Froehlich, A., and A. Siebrits. 2019. *Space supporting Africa: A primary needs approach and Africa's emerging space middle powers*. Cham: Springer.
FSD. 2020a. Foundation for space development. Available at https://developspacesa.org/.
FSD. 2020b. Foundation for space development. Available at https://www.linkedin.com/company/foundation-for-space-development/about/.
Geosmart. 2020. Bringing space down to earth. Available at https://geosmart.space/bringing-space-down-to-earth.html.
Gunter's Space Page. 2020. https://space.skyrocket.de/.
Harding, R.C. 2013. *Space policy in developing countries: The search for security and development on the final frontier*. New York: Routledge.
IAF. 2020a. Denel Spaceteq. Available at https://www.iafastro.org/membership/all-members/denel-spaceteq.html.
IAF. 2020b. South African space association (SASA). Available at https://www.iafastro.org/membership/all-members/south-african-space-association-(sasa).html.
Lal, B., Sylak-Glassman, E.J., Mineiro, M.C., Gupta, N., Pratt, L.M., Azari, A.R. 2015. Global trends in space volume 1: Background and overall findings. IDA Paper P-5242, Vol. 1 (June). Washington, DC: IDA Science & Technology Policy Institute.

Luncedo, N., and K. Gottschalk. 2013. The growth of space science in African countries for Earth observation in the 21st century. *South African Journal of Science* 109 (1–2): 1–5.

Marcom. 2020. Base. Available at https://www.marcom-as.com/.

Marnic. 2020a. About us. Available at https://marnicsa.co.za/Pages/About-Us.asp. Accessed on 20 Nov 2020.

Marnic. 2020b. Services. Available at https://marnicsa.co.za/Pages/Services.asp. Accessed on 20 Nov 2020.

Martin, G. 2012. SumbandilaSat beyond repair. DefenceWeb, January 25.

Martinez, P. 2016. The development and initial implementation of South Africa's national space policy. *Space Policy* 37: 30–38.

Mistry, D., and B. Gopalaswamy. 2012. Ballistic missiles and space launcher vehicles in regional powers. *Astropolitics* 10 (2): 126–151.

Munsami, V. 2014. South Africa's national space policy: The dawn of a new space era. *Space Policy* 30: 1–6.

Mzansisat. 2020. Profile. https://www.mzansisat.com/profile.

NACI. 2020. About NACI. Available at http://www.naci.org.za/index.php/about-naci/. Accessed on 13 Aug 2020.

National Government (RSA). 2020. Department of Communications and Digital Technologies (DCDT). Available at https://www.nationalgovernment.co.za/units/view/428/department-of-communications-and-digital-technologies-dcdt. Accessed on 15 Aug 2020.

NDT. 2020a. Home. Available at http://www.ndt.co.za/index.html. Accessed on 2 Sept 2020.

NDT. 2020b. About. Available at http://www.ndt.co.za/about.html. Accessed on 2 Sept 2020.

NRF. 2020a. Mandate. Available at https://www.nrf.ac.za/about-nrf/mandate. Accessed on 6 Aug 2020.

NRF. 2020b. Short history of the National Research Foundation. Available at https://www.nrf.ac.za/about-nrf/history. Accessed on 6 Aug 2020.

NSS. 2020. Products. https://www.newspacesystems.com/products/.

Ogunnubi, O., and S. Oyewole, eds. 2020. *Power politics in Africa: Nigeria and South Africa in comparative perspective*. Newcastle: Cambridge Scholar Publishing.

Oyewole, S. 2017. Space research and development in Africa. *Astropolitics* 15 (2): 185–208.

Oyewole, S. 2020a. The quest for space capabilities and military security in Africa. *South African Journal of International Affairs* 27 (2): 147–172.

Oyewole, S. 2023a. The African orbital journey in space age: From spectator to participant. *Vanguardia Dossier* 88 (15): 72–75.

Oyewole, S. 2023c. Defence and security policy of South Africa. In *Readings in defence, security and strategic studies*, ed. F.C. Onuoha, G.E. Ezirim, and C.M.A. Kwaja, 124–135. Nsukka: University of Nigeria Press Limited.

Oyewole, S. 2020b. Military capabilities of regional powers in Africa: Nigeria and South Africa in comparative perspective. In Ogunnubi, O., Oyewole, S. (eds.), *Power politics in Africa: Nigeria and South Africa in comparative perspective*. Newcastle: Cambridge Scholar Publishing Ltd.

Oyewole, S. 2023b. Military space strategies and African realism: Egypt, South Africa and Nigeria. In: *Oxford handbook of space security*, ed. S.M. Pekkanen, P.J. Blount. Oxford: Oxford University Press.

Pinkmatter. 2020. Who we are. Available at https://www.pinkmatter.com/. Accessed on 20 Sept 2020.

Reutech. 2020a. Company profile. Available at https://www.reutech.co.za/. Accessed on 2 Sept 2020.

Reutech. 2020b. Reutech communications. Available at http://www.reutechcomms.com/. Accessed on 2 Sept 2020.

RSA. 2008. South African National Space Agency Act. Cape Town: RSA.

SAC. 2020. Welcome to space advisory. Available at http://www.spaceadvisory.com/. Accessed on 6 Nov 2020.

SANSA. 2014. *SANSA annual report 2013/2014*. Pretoria: SANSA.

SANSA. 2015. *SANSA annual report 2014/2015*. Pretoria: SANSA.

SANSA. 2016. *SANSA annual report 2015/2016*. Pretoria: SANSA.

SANSA. 2017. *SANSA annual report 2016/2017*. Pretoria: SANSA.

SANSA. 2018a. *SANSA annual report 2017/2018*. Pretoria: SANSA.

SANSA. 2018b. SANSA actively involved in supporting floods disaster planning and awareness. *SANSA News*, February 21.

SANSA. 2019a. *SANSA annual report 2018/2019*. Pretoria: SANSA.

SANSA. 2019b. Satellite data says 2017 fires destroyed 20% of Knysna forest; caused more dangerous weather. *SANSA News*, April 26.

SANSA. 2019c. Role of satellite imagery in cyclone Idai. *SANSA News*, March 28.

SANSA. 2019d. Agricultural field campaigns underway at pilot sites in the Free State and KwaZulu-Natal. *SANSA News*, March 29.

SANSA. 2020a. *SANSA annual performance plan 2020/2021*. Pretoria: SANSA.

SANSA. 2020b. About. Available at https://www.sansa.org.za/about-sansa/. Accessed on 12 Aug 2020.

SANSA. 2020c. Facilities. Available at https://www.sansa.org.za/facilities/. Accessed on 12 Aug 2020.

Seminari, S. 2019. Global government space budgets continues multiyear rebound. *Space News*, November 24, 2019.

Sentech. 2020. Who we are. Available at https://www.sentech.co.za/about-us/who-we-are. Accessed on 5 Nov 2020.

SGAC. 2020a. SGAC. Available at https://spacegeneration.org/. Accessed on 21 June 2020.

SGAC. 2020b. SGAC South Africa. Available at https://spacegeneration.org/regions/africa/south-africa. Accessed on 21 June 2020.

Simera. 2020. More resolution. smaller satellite. Available at https://simera-sense.com/. Accessed on 11 Aug 2020.

Space in Africa. 2018. South Africa ad Nigeria supremacy on space investment, September 17.

Space in Africa. 2019a. *NewSpace Africa industry report*. Space in Africa.

Space in Africa. 2019b. *African space industry annual report* (2019 Edition). Lagos.

Space in Africa. 2019c. 2019 might be the best year so far for the African space industry, December 23.

Space in Africa. 2019d. SCS space nSight-1 exceeds expected lifespan in orbit, April 17.

Space in Africa. 2019e. First data from South Africa's ZaCube-2 to be unveiled on Tuesday, February 23.

Space in Africa. 2020d. Exclusive interview with Bryan Dean, CEO of dragonfly aerospace, October 15.

Spescom. 2020. Spescom Ltd. https://za.linkedin.com/company/spescom-limited.

Amaya Space. 2020. "Who We are", available at https://amayaspace.com/

SCS Space. 2020. About. Available at http://scs-space.com/about/. Accessed on Nov 2020.

Spaceteq. 2020. About. Available at https://www.spaceteq.co.za/. Accessed on 11 Nov 2020.

Stone Three. 2020. About. Available at https://www.stonethree.com/about/about/. Accessed on 12 Nov 2020.

Stratosat. 2020. About US. Available at https://www.stratosat.com/about-stratosat-datacom/. Accessed on 12 Nov 2020.

The Tauri Group. 2011. *The South African space industry*. Cape Town: Foundation for Space Development South Africa (FSDSA).

TIA. 2020. About technology innovation agency. Available at https://www.tia.org.za/about-us/. Accessed on 15 Aug 2020.

UCS (Union of Concerned Scientists). 2020. Satellite database. Available at: https://www.ucsusa.org/resources/satellite-database. Accessed on 28 Jan 2020.

Wener, D. 2018. Will ThinSats inspire the next generation of engineers and scientists?. *SpaceNews*, February 21.

Wood, D., and A. Weigel. 2011. Building technological capability within satellite programs in developing countries. *Acta Astronautica* 69: 1110–1122.

8 The Contribution of Space Policy to Development and Security in South Africa

Abstract

This chapter examines the extent that South Africa has employed its space and related capacities and capabilities (such as institutional, financial, human and material resources, access and assets) to advance both the national and human dimensions of development and security of the country. It provides insights into the convergence and divergence of space policy objectives of South Africa and the extent of their realisation. Accordingly, this chapter interrogates the progress and challenges of South Africa in space support for national competitiveness, economic growth and transformation, indigenisation of technological capacity, defence, security, power projection, disaster management, human rights and the struggle for freedom, democracy, good governance, food and water security, sustainable environment, quality health and education systems, and international cooperation as indicators for both (national and human) development and security.

8.1 Introduction

The growing public advocacy, policy priority, and investment, and institutional, human capital and infrastructural developments for space R&D in South Africa, as evident in the last two chapters, have generated public and academic interests in the benefits for the government and people of the country. In this case, there is a need to transcend the perspective that consider the development of space policy, strategy, programme, institution, investment, project, capacities and capabilities as developmental goals to be achieved. In the bigger picture, the development and security aspirations of the people and government of South Africa necessitate the need to examine and re-examine space and the attendant policy, investment, institutions, programmes, capacities and capabilities as means to an end. Accordingly,

S. Oyewole, *Utilitarianism in Outer Space*, Southern Space Studies,
https://doi.org/10.1007/978-3-031-49646-2_8

this chapter focuses on the extent that space and the attendant technologies have supported the development and security interests of South Africa.

Beyond this introduction and the conclusion, this chapter is broadly divided into two sections. The first of the main sections interrogates space support for national development and security objectives of South Africa. The second section provides insights into space support for human or people-centric development and security interests of the country. These are followed by a concluding section.

8.2 Space Support for National Development and Security in South Africa

As earlier noted, the space policy and programmes of South Africa were designed with the aim of promoting and supporting national interests, that is, the aspirations of the government of the country. This section examines the level of progress or the extent of achievement recorded by the country in relation to some key national interests in space. In doing this, this section is broadly subdivided into two subsections: the first one interrogates space support for national development, while the other one examines space support for national security, in South Africa.

8.2.1 Space Support for National Development in South Africa

The chapter two of this volume reviewed different themes, conceptions and perspectives on development, amidst which those that focus on national development were identified. Accordingly, the objectives of development in this case include economic growth, liberalisation and transformation, resource management, catching up or competitiveness, local or independent technological capacity, among others. These equally reflect some of the objectives of national space policy, science and technology policy, and development conceptions of the government of South Africa, as earlier identified in the chapter six of this volume. Some of these objectives of development will be examined in relation to space support under three subsections. The first subsection focuses on development as competitiveness and catching up, the second subsection focuses on development as economic growth, liberalisation and transformation, while the third subsection considers local or independent capacity and self-reliance of South Africa in space.

8.2.1.1 Development as Competitiveness and Catching up

As earlier identified, securing competitive advantage against other nations, staying comparable to peer group of nations, and catching up with advance nations, in international political economy can be considered as a perspective in the conception of development goals. This perspective is well recognised in the conception and framing of South Africa's space policy and associated end. For instance, the South African Astronomy Geography Advantage Act of 2007 that was assented to in 2008 formalised and legalised the long held believe and aspiration of the country to be a leading player from the Southern Hemisphere, and destination with competitive advantage for, astronomical R&D of southern skies. The South Africa's space policy and strategy equally emphasis the desire to develop competitive space industry that can drive international competitiveness of the country, most especially in promoting export and capturing a share of global market in this sector. In this consideration, South Africa is recognised generally and in the context of space R&D as a developing nation and emerging player globally, and regional power in Africa.

At the global level, South Africa like many other developing nations, has adopted space policy and develop considerable capacity in this area in an attempt to catch up, as well as close the development gap, with the advanced nations. From its location and aspiration in the Global-South, the country has developed considerable capacity for competitiveness and competitive advantage as well as attraction for internationally sponsored projects, especially in astronomical R&D. These are evident in the ways that the development of astronomical facilities in South Africa has benefited from the country's strategic attraction of sponsorship, collaboration, technological transfer, consulting, funding, staffing, training, and other support globally. Most of the telescopes in the country owe their existence to geostrategic competitive advantage and attraction of the country to scientific communities and their policy-makers across Europe and North America, and recently China. The success of South Africa against Australia in the bid for siting the construction of SKA, a project sponsored through international collaboration, is a testimony to the competitiveness of the country.

South Africa has equally developed considerable industrial capacity to capture a share of the global space market, especially by the Global-South and African standard. For instance, Hartbeeshoek facility offers a competitive TT&C service for clients from across the world. It has supported more than 600 spacecraft in Low-Earth orbits and high-altitude geostationary orbits (SANSA 2020a). Armscor and Denel have equally exported made in South Africa guided tactical missiles internationally. A growing number of South African companies have exported satellite subsystems and components across the world. For instance, in more than 30 years of its existence, NNS has over 60 clients, including more than 9 reseller and 14 national space agencies across the world (NNS 2020). Moreover, some South Africa's companies, especially those that are involved in communications, with a combination of terrestrial and space infrastructures, have developed into multinational giants. These among other things offer insight into the growing competitiveness of South Africa as well as position of the country in the global map of space and attendant technological market.

Despite the foregoing, South Africa's share of the global space economy is still low and less significant in the general picture of things (Euroconsultant 2017; OECD 2014; Space Foundation 2015). For instance, South Africa only accounted for 0.03% and 0.05% of global space expenditure in 2016 and 2018, respectively (Euroconsultant 2017; Seminari 2019). Besides, most of the satellite systems that were produced in South Africa largely remain in the realm of technological demonstration, although some of the subsystems have entered the commercialisation stage. More policy commitments and investment are still required to make South Africa's satellite and general space industry globally competitive and export viable. In this consideration, space and related technologies that underscored the technological hub initiative of the South African government is commendable. However, these concerns are relevant in locating the country among other emerging space players. As such, South Africa is largely considered as a second-class developing nation or emerging player in space, with few countries ahead of it in the Global-South (Harding 2013; Martinez 2016). In this case, absence of orbital launcher capability as well as the scope and size of the space policy interests, investment, industry and market are issues that affect the global rating of the RSA.

Despite the foregoing concerns against the competitiveness of South Africa at the global level and the drive to catch up with the advance nations in space, the country retains competitive advantage in many areas in Africa. The country is setting the pace in many areas of space R&D, industry and militarisation. For instance, South Africa is home to 71% of all identified astronomical facilities in the continent (AFAS 2020; Oyewole 2020a). As at the middle of 2020, the country accounted for all satellites fully manufactured within Africa by citizens of African nations. A report of Space in Africa (2019) shows that South Africa accounted for 87.5% of all known space companies in the continent. The same report shows that total number of space companies that are based in Cape Town and Stellenbosch alone are more than those in the rest of Africa. Besides, while majority of space companies across Africa only target national and regional markets, many space companies in South Africa are already in the global league. With the launch of Condor-E2 in 2014, South Africa became the first African country to sponsor, own and operate a satellite designed for military purpose. These among others show that the RSA retain competitive advantage in many areas of space R&D, commercialisation and passive militarisation, of which other African countries have to struggle to catch up.

8.2.1.2 Development as Economic Growth, Liberalisation and Transformation

The space policy and programmes of South Africa has considerably contributed to national development, especially in the areas of economic growth, liberalisation and transformation. To examine these in details, this section will be divided into two subsections. The first subsection here focuses on the contribution of space policy, institutions, programmes, investment and capabilities of South Africa to desired economic growth and liberalisation, while the second subsection examines them in relations to economic growth and transformation.

Development as Economic Growth and Liberalisation

More wealth has been created as investment in space sector has continued to increase the size and growth of the South Africa's economy in its own ways. For instance, records of SANSA between 2013/2014 and 2018/2019 financial years show that the agency's revenues always surpass its expenditures, from where surplus and not deficit of actual amount on a comparable basis were generated annually. The surplus was approximately R 101.6 million in 2013/2014, R 79.4 million in 2014/2015, R 66.1 million in 2015/2016, R 135.6 million in 2016/2017, R 72.2 million in 2017/2018 and R 130.7 million in 2018/2019 (SANSA 2014, 2015a, 2016a, 2017a, 2018a, 2019b). In most cases, between 25% and 40% of the agency's revenue are derived from non-government subsidy sources, such as contracts with public, private and foreign entities. At least, 37.6% of the agency's revenue in 2018/2019 are from such non-government subsidy sources (SANSA 2019b). With about 200 employees across its facilities, the state-owned Sentech equally generates US$ 115.56 million in sales (Dun and Bradstreet 2020). These are indicators of commercialisation and liberalisation of public entity, and their utilitarian contribution to economic growth in South Africa.

As desired by prophets of liberalism and emphasised as one of the objectives of South Africa's space policy, the country has made significant progress to make the interests, efforts and investments of private individuals and organisations a major driven force of its policy, programmes, investment and capabilities in the space sector. As evident in the chapter seven of this volume, there is a growing number of private entities in South Africa's space sector. This has considerably shared or shifted part of the burden of space R&D to the private sector, allowing government to accommodate more public interests in its list of policy priority. Besides, by reducing the influence of politics and bureaucracy in investment decision, liberalisation is believed to be more economically effective and efficient, including in space exploration and exploitation (Anderson 2013; Pelton 1992). Accordingly, privatisation and commercialisation have attracted more local and foreign investment into the space sector, and created more wealth, in South Africa.

The space-related private sector has considerably contributed to economic growth in South Africa. First, it has made it possible to retain most of the public investment in space locally. Second, it raised fund locally through stock market and other private sources as well as attract foreign investment. Although many of them did not disclose details of their revenue and investment, the report of Space in Africa (2019) offers some useful insight into the contribution of the sector to the economy through the investment raised and net worth of some of the space companies in the country and beyond. For instance, Aerobotics had raised US$ 4.6 million in total investment as of 2019. Similarly, Astrofica raised US$ 300,000 in total investment. Amaya Space and Simera Sense were reported to have secured contracts of US$ 2 million and US$ 1.5 million respectively in early 2019. Mzansisat was evaluated to worth US$ 54.7 million, while XinaBox and Newspace worth US$ 10 million and US$ 2.5 million respectively. Geosmart was evaluated to worth US$ 500,000 with close to US$ 100,000 annual revenues, while Stratosat makes between US$ 50 and 100 million annually (Space in Africa 2019).

Beyond foreign investment raised by the private sector, the government of South Africa has equally succeeded in attracting similar investment for space facility and infrastructural development, with significant implication for economic growth. For instance, South Africa only contributes a third of the US$ 35 million cost of constructing SALT, while Britain, Germany, Poland, New Zealand, and the United States paid the rest of the bill. Similarly, South Africa is only expected to pay a fraction of the cost of the ongoing multinational efforts on the construction of SKA, which is scheduled to cost about US$ 3.94 billion (Oyewole 2017a, 2020a). These are similar to the experience of the country on other major space projects.

Against this background, it is evident that space policy, programmes, and projects are largely liberalised in South Africa, with significant implications for economic growth and wealth creation in the country. Nevertheless, some observers are worried that government's role is complicated and challenging for the private sector, as policy is largely unpredictable and with less stable orientation in South Africa like other developing countries. In this case, the challenges of slow economic growth and dwindling budget cannot be ignored in government investment in space sector. Moreover, most of the existing space companies are believed to rely and survive on state-funded programmes and projects, which are marked with policy inconsistency and instability in funding. Private companies cannot invest money where there is little or no customers, while non-governmental customers only want to do business with companies that they can be sure of its continuous existence. These are the reasons why some companies have closedown, exited, or redirected their investment into other sectors. Prominent among the exited cases are Omnipless Pty Ltd., Capricorn Satellite Communication, iKubu Ltd., Maxwell Satellite Technology Pty Ltd., and 03N Networks Ltd (cf. Space in Africa 2019). As a key informant disclosed, these among other reasons have produced a trend, which only "create a new generation of frustrated space professionals every decade or two who then emigrate to work elsewhere and contribute to brain drain." In this manner, government commitment and consistency are essential for private sector survival.

Development as Economic Growth and Transformation

Space policy and attendant technologies have contributed to the economic transformation of South Africa, like many other countries across the world. First, the implementation of the space policy among other policies on science and technology, communication, industry, and commerce have recorded considerable success in creating and sustaining a space sector among other sectors that are responsible for the country's high and medium-tech industries. As of 2018, South Africa exported 5.3% of its manufactured high-tech in exchange for US$ 2.2 billion. Earlier in 2017, South Africa exported 46.7% of its manufactured high and medium tech in exchange for US$ 24.4 billion (World Bank 2020). Beyond the transformational impacts of these high-technologies, which include space-related ones, they are sources of foreign exchange, with economic contribution and prospects, in South Africa.

Second, space support is evident in the transformation of different sectors of South Africa's economy, including communication, digitalisation, urban planning, construction, resources management, and education, which have all boosted the economy and created more wealth.

Over the last three decades, South Africa has recorded progress in adopting communication technologies and digitalisation of its economy. Within this period, more internet penetration was recorded, with growing space and terrestrial infrastructural support. The importance of rapid adoption and access to mobile phone in the country cannot also be overstated. Available data from the World Bank (2020) shows that the percentage of the population with internet access in South Africa rose from 0.013 in 1991 to 6.347 in 2001, 33.97 in 2011 and 56.167 in 2017. These have revolutionised different sectors of the economy with the rise of the e-commerce, e-learning, e-governance, precision agriculture, and telemedicine. In connection to these, Mzansisat and Stratosat datacom among other space companies provide broadband services to support mining, government, transportation, enterprise, NGO, construction, defence, oil and gas, finance, health, agriculture, and ICT. As a result of these among other activities, online retail was projected to be 1.4% of the estimated R 1 trillion to be spent on traditional channels in 2018 (GCIS 2019). However, the economic transformation is not evenly shared in the country. While internet access is high in some provinces such as Gauteng (74.6%), Western Cape (72.4%) and Mpumalanga (70.2%), they remain lower in others such as Limpopo (46.2%) and Eastern Cape (55.3%). Equally, internal access in rural areas (45.0%) is low compared to metros (67.5%) and urban areas (63.7%) (GCIS 2019).

Besides internet, the impacts of satellite broadcasting for TV and radio are also revolutionary in South Africa's social-cultural, political, and economic spaces. The state-owned SABC and other entities like Multichoice, Top-TV, e.sat tv, Newzroom Afrika as well as some of the state-owned and private radios operate on satellite supported platforms. The operations of these entities, which are space supported, are critical to political and strategic communication, nation-building and integration, entertainment, businesses, job opportunities and poverty alleviation. Without a fully own communication satellite, however, South Africa has continued to pay for exclusive reliance on foreign entities, especially American and European multinational corporations, to access satellite data that it utilised for broadcasting and internet among other things. In the early years of 2000s, it was reported that various users of satellite facilities in South Africa, such as Telkom, Sentech, and Transtel, spent between R 450 million and R 550 million annually to lease satellite time (Purkitt and Burgess 2005; Oyewole 2017a).

There is also a growing resort to air and space support for agriculture, urban planning, construction, and resources management in South Africa. Accordingly, drone and satellite data are having revolutionary impacts in many sectors of the economy. It was reported in early 2015 that SANSA's Earth observation directorate revealed that the French Spot-6 and Spot-7 satellite data used by government entities, such as the departments of human settlements, agriculture, forestry and fisheries, and statistics, are purchased at a cost of R 35 million per year (Wild 2015). These data are critical to guide the decisions and strategic planning of policymakers as well as tactics and operations of civil servants and corporate organisations that implement them on agriculture, human settlement, forestry and fisheries. As a result of the growing business opportunities, some companies have developed varying capacity to deliver different space supported services in these areas. Aerobotics and GeoSmart are prominent among companies that deal with satellite imageries for various clients. It is instructive that Marnic Precision Engineering, ETSE Electronics and other companies also spread their interests from space to agriculture and mining among others.

8.2.1.3 Development as Local/Independent Capacity and Self-Reliance

The objectives of South Africa's space policy include the desire to develop and promote the capacity of the country in space R&D, industry and commerce. Although the policy priority and investment that space has received, as evident with the financial resources that are made available, are inadequate, especially in comparison with some other African countries, they have been managed in ways that have produced considerable results. At the basic level, South Africa has recorded considerable progress in the development of manpower that are required for research, design, development, manufacturing, assembling, integration, testing, patenting, commercialisation, popularisation and operation of space technologies. The country has equally witness increase in the number of facilities and associated opportunities in space R&D and industry. These among other factors are driving South Africa's progress towards local and in some cases independent capacity in some areas of space programming.

South Africa has developed considerable local capacity to access some space resources as well as covert and utilise them. With available astronomical facilities in South Africa, it has been possible for the country to advance in STEM, with a level of independence in capacity to determine the mystery of space science and contribute meaningfully to the body of related knowledge. The country has equally developed notable local capacity to access space support and convert it to advance the national interests. The level of independence that the country has achieved in production of satellite is an indicator of its advancement towards self-reliance. Besides, many public and private institutions have developed the capacity to convert or integrate space support into basic services and products that are made easily accessible to the larger population. As a consequence, space supported bandwidth and imageries as well as guidance/ navigation are provided in a competitive market in the RSA.

The most remarkable aspect of local capacity development by South Africa in space entails satellite systems and subsystems. As earlier mentioned, South Africa is the first African country and the only one until 2021 to have produced locally made satellite from scratch to finish (Oyewole 2023a). The country has eleven of such satellites to its credit (see Table 7.4). Equally, the production, commercialisation and in some cases exportation of South African satellite subsystems are major achievement in consideration of the desire of the country to develop local capacity that can produced self-reliance and international competitiveness. With the first flight of Cubespace's ADCS products in 2014, the company had delivered more than 70 ADCS units and over 250 Wheels as at late 2020. The company also boasted of having more than 40 missions in orbit using its products (Cubespace 2020). At the same time, another South African company that produced ADCS solution, NewSpace System recognised that more than 60 clients are using its products (NSS 2020). These among others shows the level of capacity that South Africa has developed locally in R&D of satellite system and subsystems. It was also learnt that a South African made satellite was silently sold to a country in the Middle East some years ago. Meanwhile, the growing capacity of the Dragonfly Aerospace has the potential to boost the production capacity of South Africa in satellite systems and subsystems.

Despite the foregoing claims on the progress made by South Africa in the development of local capacity in space and attendant technologies, the country still relies on other countries or foreign entities for many things. As earlier mentioned, the country relies on other countries for launcher services, despite its rich history and renewed interests in missile and rocket R&D. At the same time, South Africa exclusively relies on foreign satellite data providers for telecommunication and navigation. Moreover, the progress made by the country in R&D of EO satellites have not eliminated or significantly reduced its reliance on foreign providers of these services. In as much as the world has become interconnected, these are subjects of concerns for students of development studies, with background in postcolonial theory, and security studies.

8.2.2 Space Support for National Security in South Africa

South Africa's space policy, programmes and capabilities have considerably advanced its national security interests. These include in the areas of defence, power projection, law enforcement and disaster management. Although South Africa, unlike Nigeria and the USA, does not have dedicated defence and security institutions with overt space functions, its planning and implementation of space policy and strategy involve government and security departments, agencies, services, and corporations as identified in chapter seven. Consequently, the national defence and security interests and programmes of South Africa have benefited considerably from space support.

Some of the defence objectives of South Africa are to deter and prevent conflict, protect national interests, and defend the territorial integrity of the country, safeguard its borders and critical infrastructures, ensure information security, promote strategic influence as well as contribute to peace and stability in the region (DOD 2015). Space capabilities have in some ways contributed to the advancement of these objectives. Some government officials and experts considered the South African sponsored first military satellite, Condor-E2 to have contributed to promote national security interests of the country (cf. Oyewole 2020a, 2023b, c). However, others are of critical opinion against this position, on the ground that military threats against the country, especially from the frontline states and their allies, have declined or vanished since the end of apartheid regime. Amidst these are those that considered it to be an irrelevant and wasteful project that sought to advance personal ego of its promoters. These cannot be total disconnected from the controversy that surround the secrecy of the project (see Joubert 2014; Helfrich 2014).

Despite some of the critiques of the defence space project, this and other space technologies have enhanced the capacity of the RSA for deterrence, territorial defence, borders security, critical infrastructural protection, strategic influence, and contribution to peace and stability in the region. Although the threats of inter-state wars have reduced significantly, local and transnational armed non-state groups pose considerable security challenges to the country's internally, maritime domain and regional interests. The DOD and the intelligence services have been engaged in deterrence and campaigns against smuggling, transnational terrorism and their network, and other trans-border crimes. Moreover, the east and west coasts of Africa are threatened by illegal, unreported and unregulated (IUU) fishing, sea piracy and other maritime crimes, of which South Africa is playing a critical role in maintaining order at the southern seas, especially in the Indian Oceans (Daxecker and Prins 2021; Oyewole 2016, 2017b). In almost all these cases, space supported C4ISR contributed to the South Africa's defence and security postures, deterrence, and campaigns. In connection to these, there were indications that the DOD initiated a process of fitting military vehicles and other assets with satellite tracking systems (DOD 2013). Shortly after the launch of Condor-E2, it was also reported that the Joint Air Reconnaissance Intelligence Centre was upgraded with space integrated and supported ACTionable INTelligence (ACTINT) reconnaissance evaluation devices, supplied by Denel Integrated Systems & Maritime (Denel ISM) and the Airbus Defence and Space (ADS) (SAAF 2015).

As one of the leading regional powers in Africa, South Africa is an important player in regional security architecture and major military powerhouse and stabiliser in the Southern and Central Africa sub-regions (Amusan and Oyewole 2017; Ogunnubi 2014; Ogunnubi and Oyewole 2020). Accordingly, South Africa is a major contributor of military forces for capacity building of other nations, peacekeeping and peace-enforcement in the region. The country has deployed troops to Burundi, CAR, DR Congo, Ethiopia, Eritrea, Mozambique, and Sudan. It deployed a total of 334,176 military personnel for peacekeeping in Africa between 1990 and 2018 (Oyewole 2020b). In almost all these cases, space supported C4ISR contributed to the South Africa's military capabilities, strategic influence, deterrence, interventions, and campaigns. This was salient in its peacekeeping arrangement and operation in DR Congo (Oyewole 2023b).

South Africa's space and related industries are also important in strategic considerations of its defence and security postures, strategic influence, and deterrence. Denel, Armscor, Ansys, Cobham, Marnic, Reutech, Stratosat, Tellumat and Dragonfly Aerospace are prominent among companies with interests and investment in defence and space related products and services. The capacity of the country to manufacture high and medium technologies that are relevant for military space programming and defence generally, which broadly include products such as spacecraft (satellite), aircraft, Armor tank, Armor Personnel Carrier (APC), ship, radar, guided missiles among other things is a major boost to the defence posture of the nation in the region and beyond. The commercialisation of some of these products, their subsystems and related services have equally added to the geostrategic influence of the country. For instance, South Africa exported US$ 2.44

billion worth of weapons to 71 recipients, including 68 countries (31 of which are in Africa), two international organisations and one liberation movement between 1950 and 2018. This amounted to the largest record of weapon exportation by an African nation. It is more than the net worth of weapon exported by Algeria, Angola, DR Congo, Egypt, Morocco, Nigeria, and Sudan combined within the period (Oyewole 2020b).

Institutions that are responsible and central to internal and situational security situations management in South Africa have also been supported with space capabilities. The South African Police Service (SAPS) and the Directorate for Priority Crime Investigation (DPCI) or the Hawks are central to internal law enforcement. These institutions among others have drawn space support, especially utilising satellite imagery, navigation, and communication to enhance their operational capacity and delivery of policing functions in South Africa. This is important in the light of growing cybersecurity threats, xenophobic uprising, and criminal violence, such as gang wars, armed robbery, drug and human trafficking, looting, carjacking, and smuggling among others in South Africa (Amusan and Mchunu 2017; Oyewole 2023c; Tafira 2018).

Space support in further evident in South Africa's disaster management strategy as a matter of national security. The country has recorded natural, technological and other complex disasters, such as flood, storm, landslide, drought, epidemic, wildfire, transport and industrial accidents, and tsunamis among others. No fewer than 308 cases of disaster that claimed 7,358 lives were recorded in South Africa between January 1900 and July 2022. This accounted for 5.8% and 15.3% of the total disasters and associated fatalities recorded in Africa within this period, respectively (Oyewole 2022b). In some of these and other related situational security challenges, disaster management strategy and operations have benefited from space support. In many cases, satellite data is critical to early warning and real time situation awareness, which are essential strategic decisions and operational responses to natural and man-made disasters.

The operations of the National Disaster Management Centre (NDMC), which is designed for emergency management in South Africa, has been supported with space policy, programmes, and capabilities. For instance, space support was evident in South Africa's responses to tsunamis and lightning that occurred in Marion Island on September 23, 2020 (SANSA 2020b). Responses to fire disasters of Worcester in February 2016 and Knysna in June 2017 also benefited from space support (SANSA 2016b, 2019a). Space technologies have equally supported the country's responses to floods disasters locally— as evident in Gauteng in 2016, UKN and Eastern Cape in April 2019 and elsewhere (SANSA 2016c, d, 2018b, 2019c). South Africa also supported other countries with satellite data for disaster monitoring, as evident with response to floods in Malawi, Mozambique, and Madagascar in 2015 as well as Cyclone Idai in Mozambique, Malawi, and Zimbabwe in 2019 (SANSA 2015b, 2019d; Oyewole 2022b). In these and other ways, the space policy, institutions, programmes, and capabilities of South Africa have contributed to its defence, security, law enforcement and disaster management.

8.3 Space Support for Sustainable Human Development and Security in South Africa

Space support for development and security in South Africa has extended beyond the state-centric conceptions to people-centric understanding of policy objectives. Accordingly, the relevance of space and attendant technologies in South African struggle to advance, create and promote job opportunities, wealth creation, poverty alleviation, education, environment, food and water security, human rights and dignity, and disaster management cannot be ignored as a subject of human development and security. This section interrogates space support for human development and security goals, focusing on human rights, good governance, food, water, and environmental security, quality health system, quality education system, poverty alleviation, employment opportunities, and youth empowerment, as well as international cooperation.

8.3.1 Human Rights, Democratisation, and Good Governance

South Africa's space policy, programmes and capabilities have supported the struggle for human rights and freedom in the country and beyond. Satellite TVs and radios as well as (terrestrial and space supported) telephoning and internets have had considerable impacts on South Africa's socio-political space. These platforms have enhanced public access to information and empowered the people to mobilise for better political participation and representation in government. They allow individuals, especially from a remote part of the country, to swiftly share their experiences with human rights violation or threats to freedom, with limited restriction, and then received public attention, mobilisation and support for protection and possibly redress and compensation. Relevant policies have equally supported some South African communications giants like MTN and DSTV among others to rise to prominence in Africa, where they are providing services that are in turn supporting some of the platforms that enhances globalisation of human rights and freedom across the continent.

Satellite imagery and navigation are relevant for protection of human rights and freedom (Froehlich and Tăiatu 2020). Satellite imageries have become increasingly tenable and acceptable in court and securing justice in territorialised infractions and disputes. They are equally relevant to protect the fundamental rights to life of the people threatened by natural and other forms of disasters or accident, through space supported early warning, which can produce decisive effects when combine with quick mobilisation and proactive response to prevent, deter, or reduce the impact of the threats as well as rescue the affected members of the population. In this manner, the support of South Africa's air and space among other capabilities are noticeable in saving lives in flooded communities (Oyewole 2022b). Police also make use of these technologies to aid responses to crime, and thereby advance its role in protection of human rights and freedom.

Beyond the course of advancing and protecting human rights, space policy and capabilities have equally supported other elements of human freedom, democracy, and good governance in South Africa. The governance of agricultural, urban, water, food, environmental, health and security policy processes, especially in terms of advocacy, planning, implementation, and assessment are supported by space capabilities. While some of these points will receive more attention in subsequent sections, the role of space support for urban governance is emphasise here.

Urban governance focus on town planning with details like infrastructural development, housing, transportation, and waters. Amidst these, satellite data are playing notable roles in town planning and assessment of infrastructural development in major cities in South Africa. For instance, in the middle of 2020, SANSA sponsored a competitive modelling of location of informal settlement in KZN, using AI on the SPOT satellite imagery supplied by the agency. The competition attracted the participation of 184 scientists from across Africa, from where the winner won US$ 1,000 (SANSA 2020c). In addition to air and sea transportation, navigation satellites have further gained currency among road users and are supporting in regulation of traffic in the country. In these and other ways, some of which will be discussed later in this chapter, space policy and capabilities are supporting policy R&D to enhance governance and public services in delivering public goods that are critical to human development and security.

8.3.2 Food, Water, and Environmental Security

Space institutions, programmes, investment, and capabilities have supported research, policy and development in food, water, and environmental security in South Africa. For the purpose of food security, SANSA has engaged relevant government departments, most especially the DEFF and the DALRRD, by supplying them satellite data to make informed decisions on agriculture, land reform, forestry and fishing. The agency has also collaborated with private sector and NGOs to support farmers with space capabilities. In early 2017, for instance, the data from Sentinel-2B satellite, which was launched in 2015, and associated analysis was recognised by SANSA to have contributed significantly to policy decisions on agriculture and food security in the country (SANSA 2017b). Equally in early 2019, the agency joined other organisations in the implementation process of the African AgriCultural Systems with the support of Remote Sensing (AfriCultuRES) in Free State and KZN (SANSA 2019e). Private service providers like Aerobotics and Goesmart are further involved in air and space support for precision agriculture to enhance food production in South Africa. The growth of e-commerce, with the support of space and terrestrial infrastructures, has also aided the purchase of agricultural inputes by farmers, and the distribution and accessibility of food products, thereby advancing human development and security.

With many of the ingredients of precision agriculture imbibed, including elements of space technologies, it is instructive to note that South Africa is the foremost African country in food security consideration (Amusan and Oyewole 2023). Out of 100, South Africa's rating in food security rose from 41.44 in 1960 to 79.57 in 1990 and all time height of 126.7 in 2014 (World Bank 2020). As at 2014, food was 10.9959% of the mechanised exports, and 6.28% of mechanised imports of the country. The value added of agriculture, forestry and fishing for South Africa's economy rose from approximately US$ 807.8 million in 1960 to US$ 4.9 billion in 1990 and peaked at US$ 9.5 billion in 2011, and was US$ 8 billion in 2018 (World Bank 2020). Meanwhile, the growth of the sector is not comparable to other sectors like industry and services. Notably, the contribution of the agriculture, forestry, and fishing sector to the GDP of the country declined from 11% in 1960 to 4.2% in 1990 and 2.1% in 2018. Besides, the percentage of the population that is employed in the sector declined from 9.3% to 3.6% between 1991 and 2019. However, the country was able to raise value added per worker in the sector from US$ 6,173 in 2000 to its peak of US$ 14,433 in 2014, and was US$ 11,816 in 2018 (World Bank 2020). These, among others, underscored the significant modernisation of the sector with various precision technologies, which include elements of space strategy.

The food security situation of South Africa cannot be examined in isolation. It is connected to water and environmental security. The total annual freshwater withdrawal in South Africa rose from 12.73 billion cubic meters in 2000 to 15.5 billion cubic meters in 2013. The total annual withdrawal was 28.4% and 34.6% of the internal resources in 2000 and 2014, respectively. In 1990, 71.98%, 17.14%, and 10.88% of the total freshwater withdrawal in South Africa were for agricultural, domestic and industrial purposes respectively. As of 2013, agriculture accounted for 62.52% of the total freshwater withdrawal, while 27% and 10.48% remaining were for domestic and industrial purposes respectively (World Bank 2020). These call for careful management and policy governance of freshwater withdrawal in the country. In this context, satellite data has been utilised to support research on the quantity and quality of water that is available across South Africa. For instance, in 2014/2015, Spot-6 satellite data was used in research to determine the quality of Vaal Dam, with emphasis on the concentration of Algal bloom (SANSA 2015c). With advent of AI like the Mzansi Amanzi tool, SANSA (2020d) recognised South Africa's capacity to monitor and keep track of dams with satellite data rose from 250 to 30,000. These among others are supporting the country in the management and governance of freshwater for human development and security.

Environmental security is essential to food and water security. Environmental research and policy in South Africa have considerably benefited from space support. Satellite data are playing considerable roles in diagnosing and assessing the extent of ecological challenges face by South Africa, like many other countries (Froehlich 2019; Oyewole 2022a). For instance, in mid-2020, the agency supplied the satellite data that supported a study of air quality in the country. As such, the TROPOspheric Monitoring Instrument (TROPOMI) on board of Sentinel-5P satellite, which was launched in 2017, was used to examine the concentration of sulfur dioxide (SO_2) and nitrogen dioxide (NO_2) gases over the country (SANSA 2020e). The agency has also supported relevant departments and agencies on land, urban planning, agriculture, water, maritime affairs, fishing, forestry and disaster management with relevant satellite data, training, and other forms of capacity building to aid effective environment management. The efforts of these and other government agencies and departments have equally found the support of the private sectors and the academic community very critical in this process.

Despite the foregoing claims on the progress and prospects of space support for food, water and environmental security in South Africa, there are still some notable gaps between policy strategies and realities. In many cases, air and ground instruments are more considerable, accessible, affordable and utilised for the purposes: policy formulation, implementation, evaluation and consideration on agriculture, food, water and the environment. These alternative options are in most cases considered to be cheaper and easier to access and utilised. In addition, the knowledge of space, associated technologies and their utilities remain limited in the country. The interests of the general public in space and its relevance are mostly considered to be low like many other countries, while the interests of the academics and private sectors are between modest and low. Some observers equally believed that many of the scientists in space studies and related fields are not really in touch with holistic policy and strategic interests and realities.

8.3.3 Quality of Health System

South Africa space and related policy, institutions, programmes, investment, and capabilities have contributed considerably to the advancement of the quality of health system in the country. As applicable to many other countries, satellite TVs and Radios as well as internet have expanded the quantity and quality of health system and relevant information that are available for people to live healthy. Space has supported the terrestrial infrastructures to raise the prospects of telemedicine in South Africa. This is most pronounced in the area of bandwidth that provide those in the remote parts of the country the opportunity to locate, connect and access better medical facilities and experts that are based in the cities easily. Equally, e-commerce makes relevant drugs and other medical products in the cities more accessible for those in the remote locations. Generally, telecommunication, which is enhanced through space support, expand accessibility to medical insurance and ambulance services among other things in South Africa.

Space support for health research and policy are also critical to improve the quality of health system in South Africa. This is mostly noticeable with the use of satellite data to support policy research, decisions, and implementation during Covid-19. SANSA supplied relevant satellite data to different department and agencies of government, private sector, and research institutions to support their responses to Covid-19 and associated challenges. Satellite data were used to support contact tracing initiatives, in order to identify the location of infected persons, test and monitor those in their proximity, and contain further spread of the virus. Space support was employed to monitor lockdown compliance, as movement of vehicles along all major routes in the country were equally observed from satellite.

In similar manner, trans-border movements were also monitored with space support during this period. Satellite data were also important to location and mapping of warehouses and distribution of relief materials to different areas. They were equally utilised for researching the impact of Covid-19 lockdown on the environment, and the effects of the weather and climate on health (see SANSA 2020f, g, h).

8.3.4 Quality Education for All

South Africa's space policy, institutions, programmes, and capabilities have supported the quality of education and the quantity of those that have access to it in the country. Terrestrial and space infrastructures have supported e-learning through internet, TV, and Radio. Public education has been sponsored by the government, NGOs, private sector as well as academic and research institutions through these platforms over the years. These have expanded the quantity of people that have access to quality education in South Africa. During the Covid19 crisis, e-learning platforms were further deployed for basic and advance school educations.

STEM educations have been greatly enhanced by space policy, institutions, programmes, investments, and projects in South Africa. In 2019, for instance, SANSA engaged more than 10,000 secondary school students across the country in capacity building in the area of space application, including in GPS navigation training, to advance STEM educations (SANSA 2020i). More importantly, the agency and the DST among other government departments and agencies have partnered and supported academic and research institutions, and some cases, high schools, in R&D of satellite systems and its subsystems or components. Similar partnership and support are noticeable in R&D of astronomical instruments and rocketry, to the advancement of STEM educations in the country. This was the way and manner that the country developed eleven satellites locally and created a space industrial sector. Besides, the data obtained from the locally developed satellites, those purchased and made available by SANSA and the open sources have further advanced R&D in the fields of science, which has further better the quality of education. In relation to these, Table 8.1 shows the contributions and positions of South Africa in the metrics of knowledge production globally and regionally in selected fields of space S&T.

As evident in Scimago ranking, South Africa contributed 1,245 scholarly publications that attracted 13,173 citations to knowledge on aerospace engineering between 1996 and 2019, making the country 1st out of 43 African countries and 36th of 187 countries globally. In Astronomy and Astrophysics, South Africa contributed 5,183 scholarly publications, which attracted 145,428 citations, to the body of knowledge, making the country 1st of the 38 in regionally and 24th of 177 globally. South Africa contributed 1,838 scholarly publications with 45,615 citations to knowledge on Atmospheric Science, making the country 1st out of 53 in Africa and 29th out of 217 globally. In the field of Control and System Engineering, the country was the 4th out of 47 regionally and 42nd out of 188 globally, with 3,388

Table 8.1 South Africa's contribution to knowledge on space science and technologies, 1996–2019

Fields	No. of publications	Citations	Ranking in Africa	Ranking globally
Aerospace engineering	1,245	13,173	1/43	36/187
Astronomy and astrophysics	5,183	145,428	1/38	24/177
Atmospheric science	1,838	45,615	1/53	29/217
Control and system engineering	3,388	40,532	4/47	42/188
Global and planetary change	1,278	35,192	1/55	23/217
Physics and astronomy (miscellaneous)	6,893	108,935	1/55	42/206
Space and planetary science	5,483	163,783	1/48	27/205

Source Compiled by the author from Scimago (2020) database

scholarly publications that attracted 40,532 citations. With 1,278 scholarly publications that fetched 35,192 citations in the field of Global and Planetary Change, South Africa was the 1st out of 55 countries regionally and 42nd out of 217 countries globally. The country was ranked first in Africa and 42nd in the world for its contribution of 6,893 scholarly publications that have attracted 108,938 citations in the field of Physics and Astronomy (miscellaneous). In the field of Space and Planetary Science, South Africa contributed 5,483 publications that attracted 163,783 citations, and was ranked 1st out of 48 countries in Africa and 27th out of 205 countries globally.

Despite the foregoing claims on space support for education, there are enduring concerns that the knowledge of space, its attendant technologies and their utilisation are limited in South Africa. This is evident in a survey of expert opinion, which shows that public and to an extent academic attention for space policy and education are inadequate. In this case, the attention of human, social and business scientists, and particularly political, policy, strategic, economic and business analysts, on the subject matter fall short of the general capacities of the country.

8.3.5 Poverty Alleviation and Employment Opportunities

Space support for research, policy and different sectors of the economy has directly and indirectly affected poverty situation and employment opportunities in South Africa. As shown in chapter seven of this volume, the space industry directly employed about 2,000 persons, and many more indirectly, in South Africa. Besides, space supported telecommunication, most especially satellite TVs and Radios as well as internets, have created a chain of employment opportunities in the country like many others across the world. Space supported agriculture, food and water security, urban governance, environmental management, telemedicine, e-commerce, e-learning, e-banking, and e-governance among others have further created job opportunities for skilled workers, as well as increase effectiveness, per capital productivity and the standard of living of the affected persons. This is evident with how these and other technologies have revolutionised agriculture, and then raised the value added per worker in the sector from US$ 6,173 in 2000 to US$ 14,433 in 2014 as earlier identified (World Bank 2020).

Despite the prospects of space and the attendant technologies for poverty alleviation, there is an emerging concern that the job opportunities in space sector remain inadequate in the country, given the generally limited capacities of the government and industry. It is also worrisome that job security and opportunities, especially for the unskilled and semi-skilled South African labours will be affected negatively with the growing technological revolution, including space support and resorts to AI. This is evident in the earlier identified decline in the percentage of the population that are employed in agricultural sector, while the productivity of the sector continues to rise. This is a potentially alarming trend for a country that is already threatened by growing unemployment rate, as evident with some of the underpinning causes of xenophobia.

8.3.6 International Cooperation

The international dimension of South Africa's space policy goes beyond competitiveness of the nation against others to accommodate cooperation and collaboration. In some ways, South Africa has institutionalised its desire and support for international cooperation with membership of relevant multilateral arrangement, especially IGOs, in space affairs. Prominently, the country is one of the leading advocates of regional and global space cooperation. Accordingly, South Africa is a member of the ARM, which was established in 2003 with Algeria, Kenya, and Nigeria (Mostert and Jacobs 2008). It is one of the over 40 members of the RASCOM. The country was one the leading advocates of African Union's sponsored ASA and its founding members. It is also a member of UNOOSA. On these bases, South Africa has benefited and sponsored international assistance, partnership and support in space and attendant technologies.

The space policy, programmes, capabilities, and investment in South Africa have benefited from the support of many foreign governments, industries as well as academic and research institutions. This is evident with technical and financial support that the country received for the construction of most of its astronomical observatories. Accordingly, HartRao and some of its infrastructures benefited from South Africa's collaboration with the US and France. Two-third of the US$ 35 million cost of SALT was paid for by Britain, Germany, Poland, New Zealand, and the United States (Oyewole 2017a). The over three billion US dollar's SKA is currently funded by several countries, including Australia, Canada, China, France, Germany, Japan, India, Italy, Netherlands, Portugal, the UK, South Africa, South Korea, Spain, Sweden, and Switzerland.

South African satellite programmes and industry have also benefited from the international collaboration of the country. It is instructive that South Africa relied on purchase and open access data from satellites sponsored, launched, maintained, and operated in Europe and America. Intelsat, Spot and Sentinel satellites and supporting infrastructures by ADS are prominent in this case. The military satellite of the country was also designed, developed, and launched by Russia. South Africa relied on the US and Russia for launch services. Moreover, some of the industries that support indigenous capacity of the country in space sector are foreign investment, while others survived on transnational operations and exportation of satellite subsystems and components. As such, Aerobotics, Brolaz Projects, Convergence Partners, Pinkmatter Solutions and Stratosat datacom are prominent MNCs that are involved in space sector in South Africa. CubeSpace, NSS and SAC among other space companies in South Africa have also resorted to exportation. These allow them to add the benefits of foreign trade and investment to their survival, even while contributing and supporting other countries in space.

Many countries have benefited from South Africa's space programmes as well, especially in Africa. SALT and some other astronomical observatories liked space related programmes in universities and research centres in the country are considerably open to foreigners to develop their capacity and utilise. South Africa also provided the leadership that made it possible for many African countries to benefit from the siting of SKA extended facilities. The beneficiaries include Botswana, Ghana, Kenya, Madagascar, Mauritius, Mozambique, Namibia, and Zambia. South Africa has further supported disaster management in other African countries, especially in the Southern Africa, with relevant satellite data. These are in addition to the opportunities for trade and technological transfer in the areas of space R&D, especially satellite system or subsystems/components and guided tactical missiles, from South Africa to other countries in continent and beyond. These among other things show the desire and commitments of South Africa to cooperation among nations, especially in Africa, on space technologies.

8.4 Conclusion

This chapter contribute to existing knowledge on space policy and strategy in South Africa, as it goes beyond emphasis on history, institutions, targeted objectives and programmes of the country in space to interrogates some of the utilities that have been derived from the final frontier by the affected people and government. In this manner, this chapter transcends the globalised space popularisation project of the advance space players, as it offers insights into the benefits of space policy priority and investment based on the perspectives of the developing nations of the Global-South and the experiences of an emerging space players, such as South Africa.

This chapter offers relevant insights into the contribution of space and the attendant policy, strategy, investment, capacities and capabilities to the national competitiveness, economic growth and transformation, technological capacity, defence, security, power projection, disaster management, human rights and freedom, democracy and good governance, food and water security, sustainable environment, quality of health and education systems, and international cooperation in South Africa. In this process, this chapter accounted for the progress made by South Africa in extracting utilities from its national space policy and strategy, in the form of development and security for the government and people, its inadequacies and the challenges.

References

African Astronomical Society (AfAS). 2020. African observatories. https://africanastronomicalsociety.org.

Amusan, L., and S. Mchunu. 2017. An assessment of xenophobic/afrophobic attacks in South Africa (2008–2015): whither Batho Pele and ubuntu principles? *South African Review of Sociology* 48 (4): 1–18.

Amusan, L., and S. Oyewole. 2017. The quest for hegemony and the future of African solution to African development problems: Lessons from headways in African security sector. *Journal of Asian and African Studies* 52 (1): 21–33.

Amusan, L., and S. Oyewole. 2023. Precision agriculture and the prospects of space strategy for food security in Africa. *African Journal of Science, Technology, Innovation and Development* 15 (3): 325–336.

Anderson, C. 2013. Rethinking public–private space travel. *Space Policy* 29 (4): 266–271.

Cubespace. 2020. Overview. Available at https://www.cubespace.co.za/about/

Daxecker, U., and B. Prins. 2021. *Pirate lands: Governance and maritime piracy*. Oxford: Oxford University Press.

DOD. 2013. *Annual report 2012/2013*. Pretoria: DOD.

DOD. 2015. *Annual report 2014/2015*. Pretoria: DOD.

Dun & Bradstreet. 2020. Sentech. Available at https://www.dnb.com/business-directory/company-profiles.sentech.ea8c89245d2a2b05699534ea7b191cc2.html.

Euroconsultant. 2017. Government spending in space programs reaches $62 billion in 2016, May 30.

Froehlich, A. 2019. *Embedding space in African society: The united nations sustainable development goals 2030 supported by space applications*. Cham: Springer.

Froehlich, A., and C.M. Tăiatu. 2020. *Space in support of human rights*. Cham: Springer.

GCIS. 2019. *South Africa yearbook 2018/19* (26th edn). Pretoria: Government Communications (GCIS).

Harding, R.C. 2013. *Space policy in developing countries: The search for security and development on the final frontier*. New York: Routledge.

Helfrich, K. 2014. Defence minister enters the spy satellite fray. *DefenceWeb*, January 23.

Joubert, J. 2014. Scandal of R1.2 billion missing SA satellite. *Time*, January 19.

Martinez, P. 2016. The development and initial implementation of South Africa's national space policy. *Space Policy* 37: 30–38.

Mostert, S., and M. Jacobs. 2008. ARM constellation—establishing a regional remote sensing asset. *Acta Astronautica* 63: 221–227.

NSS. 2020. Products. https://www.newspacesystems.com/products/.

OECD. 2014. *Space economy at a glance 2014*. OECD Publishing.

Ogunnubi, O., and S. Oyewole, eds. 2020. *Power politics in Africa: Nigeria and South Africa in comparative perspective*. Newcastle: Cambridge Scholar Publishing.

Ogunnubi, O. 2014. Hegemonic order and regional stability in sub-Saharan Africa: A comparative study of Nigeria and South Africa. PhD thesis, University of KwaZulu-Natal, South Africa.

Oyewole, S. 2016. Suppressing maritime piracy in the Gulf of Guinea: The prospects and challenges of the regional players. *Australian Journal of Maritime and Ocean Affairs* 8 (2): 132–146.

Oyewole, S. 2017a. Space research and development in Africa. *Astropolitics* 15 (2): 185–208.

Oyewole, S. 2017b. Too early to celebrate! The decline of Somali piracy off the East Coast of Africa. *Strategic Analysis* 41 (6): 578–582.

Oyewole, S. 2020a. The quest for space capabilities and military security in Africa. *South African Journal of International Affairs* 27 (2): 147–172.

Oyewole, S. 2023a. The African orbital journey in space age: From spectator to participant. *Vanguardia Dossier* 88 (15): 72–75.

Oyewole, S. 2023c. Defence and security policy of South Africa. In *Readings in defence, security and strategic studies*, ed. F.C. Onuoha, G.E. Ezirim, and C.M.A. Kwaja, 124–135. Nsukka: University of Nigeria Press Limited.

Oyewole, S. 2020b. Military capabilities of regional powers in Africa: Nigeria and South Africa in comparative perspective. In *Power politics in Africa: Nigeria and South Africa in comparative perspective*, ed. Ogunnubi, Olusola and Samuel Oyewole. Newcastle: Cambridge Scholar Publishing Ltd.

Oyewole, S. 2022a. How space research can help Africa meet the SDGs. Presentation at a webinar of the University of Johannesburg's Institute for Pan-African Thought and Conversation (IPATC) on "Reaching for the final frontier: The African Space Sector", September 29.

Oyewole, S. 2022b. Air and space strategies in African disaster management. Virtual Lecture of the (British) Royal Air Force (RAF) Museum on August 11.

Oyewole, S. 2023b. Military space strategies and African realism: Egypt, South Africa and Nigeria. In *Oxford handbook of space security*, ed. S. M. Pekkanen and P. J. Blount. Oxford: Oxford University Press.

Pelton, J. N. 1992. Organizing large space activities: Why the private sector model usually wins. *Space Policy* 233–244.

Purkitt, H.E., and S.F. Burgess. 2005. *South Africa's weapons of mass destruction*. Bloomington: Indiana University Press.

SAAF. 2015. Denel ISM upgrading joint air reconnaissance intelligence centre, May 22.

SANSA. 2014. *SANSA annual report 2013/2014*. Pretoria: SANSA.

SANSA. 2015a. *SANSA annual report 2014/2015*. Pretoria: SANSA.

SANSA. 2015b. Satellite technology responding to flood disasters in Southern Africa (Malawi & Madagascar). *SANSA News*, January 27.

SANSA. 2015c. Using imagery to monitor water quality in the Vaal Dam. *SANSA News*, May 20.

SANSA. 2016a. *SANSA annual report 2015/2016*. Pretoria: SANSA.

SANSA. 2016b. Landsat 8 aiding in fire investigations in South Africa. *SANSA News*, September 7.

SANSA. 2016c. Monitoring floods in Gauteng and other parts of the country. *SANSA News*, November 15.

SANSA. 2016d. Space technology in flood prediction. *SANSA News*, November 15.

SANSA. 2017a. *SANSA annual report 2016/2017*. Pretoria: SANSA.

SANSA. 2017b. Sentinel-2B for enhanced decision-making in agriculture and food security. *SANSA News*, March 10.

SANSA. 2018a. *SANSA annual report 2017/2018*. Pretoria: SANSA.

SANSA. 2018b. SANSA actively involved in supporting floods disaster planning and awareness. *SANSA News*, February 21.

SANSA. 2019a. Satellite data says 2017 fires destroyed 20% of Knysna forest; caused more dangerous weather. *SANSA News*, April 26.

SANSA. 2019b. *SANSA annual report 2018/2019*. Pretoria: SANSA.

SANSA. 2019c. Update on recent flooding in KZN and Eastern Cape using satellite imagery. *SANSA News*, April 29.

SANSA. 2019d. Role of satellite imagery in cyclone Idai. *SANSA News*, March 28.

SANSA. 2019e. Agricultural field campaigns underway at pilot sites in the Free State and KwaZulu-Natal. *SANSA News*, March 29.

SANSA. 2020a. About. Available at https://www.sansa.org.za/about-sansa/. Accessed on 12 Aug 2020.

SANSA. 2020b. Marion Island sentinel for tsunamis and lightning. *SANSA News*, November 3.

SANSA. 2020c. African data scientists harness AI to find informal settlements in SANSA's satellite imagery. *SANSA News*, November 3.

SANSA. 2020d. Mzansi Amanzi an oasis of water management data. *SANSA News*, November 3, 2020.

SANSA. 2020e. Tracing trends in South Africa's air quality. *SANSA News*, July 22.

SANSA. 2020f. SANSA satellite data boosts government's response to COVID-19. *SANSA News*, July 22.

SANSA. 2020g. Digital earth Africa releases analysis-ready satellite data for all of Africa that can support COVID-19 response and recovery. *SANSA News*, June 19.

SANSA. 2020h. Space solutions support Government's COVID-19 containment. *SANSA News*, April 7.

SANSA. 2020i. Over 10 000 learners across South Africa saw value of space science in 2019. *SANSA News*, March 20.

Scimago. 2020. Scimago country ranking dataset, 1996–2019. https://www.scimagojr.com/countryrank.php.

Seminari, S. 2019. Global government space budgets continues multiyear rebound. *Space News*, November 24, 2019.

Space Foundation. 2015b. *The space report 2015: The Authoritative guide to space activities*. Colorado Spring: Space Foundation.

Space in Africa. 2019a. *NewSpace Africa industry report*. Space in Africa.

Tafira, H. K. 2018. *Xenophobia in South Africa: A history*. Palgrave Macmillan

Wild, S. 2015. South Africa wants more eyes in the sky. *Mybroadband*, April 3.

World Bank. 2020. World development indicators. Available at http://datatopics.worldbank.org/world-development-indicators/. Accessed 10 March 2021.

Part IV

Final Conclusions

9 Conclusion: On Space Policy, Development and Security in Nigeria and South Africa

Abstract

This chapter provides a concluding reflection on space policy, development and security in Nigeria and South Africa. It summarises major themes that featured in the subject of discussion, from where a comparative analysis of the experiences and perspectives of Nigeria and South Africa were drawn. Nigeria's space policy revolves around government and academic institutions with limited private sector, unlike South Africa's academic and private sector-driven and government coordinated model. The chapter also explores the state of knowledge on space policy and politics in the study areas, which are located in the contexts of emerging African and Global-South studies, as well as the unique contribution of this volume to literature, and relevant areas for future research.

9.1 Advancing the Frontier of Knowledge

This volume contributes to the existing body knowledge on space politics, policy and strategy, development and security, and international relations, based on the experiences and perspectives of Nigeria and South Africa. It allows for a closer encounter with, and detailed reflection on, the underrepresented and underestimated experiences, perspectives, and narratives of the Global-South, developing nations and emerging space players at the global level, and regional powers, including those in Africa, such as Nigeria and South Africa. Hence, this volume is a contribution to the efforts to address the marginalisation of African countries and others from the Global-South in knowledge production in space politics and policy, security, development, and international studies (Harding 2013; Ogunnubi and Oyewole 2020; Oyewole 2021, 2023a; Smith 2015).

It is important to note that there is a growing research attention on space policy in Africa and the Global-South in the last decade. At the global level, Harding (2013) provides a remarkable analysis on space policy in developing countries,

S. Oyewole, *Utilitarianism in Outer Space*, Southern Space Studies,
https://doi.org/10.1007/978-3-031-49646-2_9

as a departure from superpowers and developed countries-centred narratives. In Africa, national and regional review and survey of space interests, policies, strategies, institutions, programmes and historical trajectories dominated the first two decades of its orbital journey (Abiodun 2008, 2017; Akinyede and Boroffice 2013; Gottschalk 2010; Martinez 2016; Munsami 2014; van Wyk 2009). In recent years, however, there is a growing attention for African space capacities and capabilities, and the utilities (Froehlich 2019; Froehlich et al. 2020; Froehlich and Siebrits 2019; Oyewole 2017, 2020a, b, c; 2022a, b; 2023b). Prominently, a series on Space Support in Africa by Annette Froehlich, which have further encouraged the emergence of her edited Springer published Southern Space Series is important in this context.

Despite the emerging research attention on the subject in the region, the in-depth country-focus analysis in this volume, offers a unique perspective on space policy in study areas, which provide insights into African experiences and perspectives. Notably, this volume paid attention to the underexplored convergence and divergence between what Nigeria and South Africa sought to achieve through space policy and their records of achievement. It interrogates dynamics of space politics, policy, strategy, programmes, institutions, investments and capabilities of study areas, and their utilitarian contribution to development and security aspirations of the peoples and governments.

The diversity in disciplinary, theoretical, and conceptual orientations of this volume provide this analysis with the famous opportunities of "few cases, many variables" (see Porta 2008). This allows this volume to provide bridges across disciplines, especially between pure and applied sciences on one hand and art, humanities, and social sciences on the other hand. It transcends the prevailing academic focus on space, the attendant technologies and its applications in Africa to accommodate the political, policy, strategic, socioeconomic, security, and International Relations frameworks. The broad framework of the theory adopted equally made it possible to work on different conceptual perspectives and themes with several indicators on space policy (programmes, investments, human and material resources, among other capabilities) and their relevance to development and security, which ranged from state-centric to people-centric goals. With this broad scope of analysis, this volume offers a unique view with some fresh, refreshing, and integrated perspectives on space policy and politics in the selected countries, Africa and the developing world.

9.2 Summary and Conclusion

On a general note, both Nigeria and South Africa have made significant progress in space R&D and associated policy. It is remarkable that Nigeria has prioritised and recorded progress in state institutional, human, and infrastructural capacity development to operationalise its space policy aspirations over the last three decades. Besides, this volume shows progress and challenges in how Nigeria had employed space technologies to support economic growth and transformation, internal security, defence and power projection, disaster and election management, governance, health and education systems, human freedom, food, water and environmental security, and international cooperation. At the same time, South Africa has prioritised and recorded progress with state institutional, human, and infrastructural capacity development for academics and private sector to play a leading role in operationalising its space policy and strategy. It is therefore remarkable that this country has developed impressive capacity to produce and commercialise some space-related technologies as well as support its national and human development and security aspirations.

In comparative perspective, it is observed that the space policy and strategy of Nigeria are largely designed and developed around applications of space technologies by the national government and universities with limited capacity in production and commercialisation. In South Africa, however, the space policy and strategy revolve around production and commercialisation of some relevant technologies by the academics and private sectors, with applications by state and non-state entities. In another words, pattern of space institutional development in Nigeria is largely government driven with academic support and growing military dimension, with limited private sector. For South Africa, however, the institution is largely academic and private sector driven with the government support and covert military dimension. These are important for the development paths of these countries in terms of space-related human and material resource management. Compared to South Africa, Nigeria's space programmes are observed to be characterised by sophisticated but fewer and foreign acquired space assets, and less sophisticated but growing number of ground facilities and infrastructures, which are maintained with more human and financial resources that are mostly under-utilised as inputes for space capabilities and desired development and security. These development paths are models that are likely to reflect on many other African countries.

Despite their record of progress, this volume raised a series of concerns on the space policy, its operationalisation and utilisation in Nigeria and South Africa. For instance, there were concerns on space-related human resource management, especially on the quantity and quality of recruited personnel in relevant state institutions, as well as financial mis/management, in Nigeria. The underutilisation of Nigeria's human and material resources for space-related development and security is also worrisome. This is evident in the underdevelopment of the capacity of the country in design, production, and commercialisation of space technologies as well their inadequate application for law enforcement, economic transformation, urban planning, forest and maritime governance, water and environmental management, precision agriculture and food security. Besides, the potential implications of space support for the armed forces and other security forces with the enduring record of militarisation of public space have become a subject of concern for human rights and democratisation, despite the growing resort to the similar space support in the struggles for human and environmental freedom, in Nigeria (cf Oyewole 2020a, b, c).

South Africa's space policy is not without its challenges as well. There are concerns that the government has maintained inconsistent or failed to sustain policy priority and investment for space R&D. This concern is also notable in the case of Nigeria, where relevant institutions are often left handicap and human resources are made redundant due to unavailability of funds. In the case of South Africa, however, this development has forced many promising space-related high-tech companies to liquidate. In both cases, many bright minds that made up the highly skilled human resources of these countries have been forced to emigrate, thereby contributing to brain drain, a pattern of unequal and unfavourable labour mobility from the Global-South to the Global-North. Nevertheless, the limited economic growth and in some cases stagnation and contrast of the aforementioned countries in recent years have caused budget constrains with implication for space policy and investment. These among other things have equally caused underutilisation of human and in some cases material resources that are available for South Africa to realise its potential in space R&D and their application for the advancement country's development and security goals.

The covert nature of South Africa's military space programmes constitutes another subject of concerns for transparency and regional security. This is important in the light of the absence of a coherent institutional framework for the defence space programme and the shadiness of the military satellite project of the country. In a series of attempts to absorbed themselves from wrongdoing and public scrutiny, almost all the institutions that are involved in Condor-E2 or Flute project were engaged in finger-pointing and name calling of others as responsible (see Joubert 2014; Helfrich 2014). With these and related developments, the exact value of the project remains in the realm of politicised debate in South Africa. Besides, the military satellite is considered as the main catalyst for the second and emerging wave of military space programming in Africa since the 2010s (Oyewole 2020a). Without any pronounced connection, it is instructive that Nigeria established the DSA to institutionalise its military space programme in 2014, the same year that information about South Africa's secrete military satellite was leaked and launched. As at late 2020, Egypt and Morocco are observed to have acquired space assets for military or dual purposes, and Nigeria join their rank with the launch of its military satellite in early 2023.

Nigeria and South Africa, like other leading African countries that maintains some sort of space programmes and capabilities, have displayed their commitments to international cooperation in space exploration and exploitation at the global, regional and sub-regional levels. However, the drive for space capabilities for military-security has the potentials to generate competition among these countries and other regional powers in Africa. Although the potential for direct hostility is very low, this may militarise space policy and programmes for political and security purposes, in some cases at the expense of relevant development objectives. Besides, the quest for competitive advantage and catching up with advance nations or peers in security and political economic realms remain another concern for international cooperation between Nigeria and South Africa, and in the region generally. It is not surprising therefore that non-African space-related products and services are more patronised at the expense of African ones in the region. This is evident from the pattern of space-related procurements, such as satellite systems, subsystems, components, licensing, services, and data, and supporting ground based equipment purchased by many African countries.

It is against this background that Nigeria and South Africa are creating development models and paths in space policy and politics with the prospects that many other African countries will consider, follow, adopt, adapt, refine, internalise, and re-popularise. Moreover, it is important to note that the space policy-decisions, programmes, institutions, investments, and capabilities of these countries are essential components of the experiences, perspectives, and narratives of Africa, developing countries, and emerging nations and players in space politics and policy studies.

9.3 Recommendations

In response to the challenges and prospects of space policy, programmes and investment that have been examined, this volume calls for more cautious research, development, and security-oriented attention on the subject matter. In consideration of academic interests to gain and advance knowledge, there is a need for more research attention on space studies from all fields, especially in study areas and the region. In addition to the efforts of those in pure and applied sciences, more space-related research in art, humanity and social sciences are required to document the records of these nations among others as contributions to knowledge on the experiences, perspectives, and narratives of Africa by African, as well as developing countries and emerging players in space. As a member of social sciences, students of political science, economy, International Relations, and policy studies need to embrace multidisciplinary perspectives and multiple theoretical frameworks to enrich studies in space politics, policy, programmes, investments, and capabilities as well as their prospects and challenges for national, regional and human development and security interests.

The volume observes the need to balance attention for national-centred space policy interests and programmes with the regional-centred ones in Africa, including in analysis. There is a need to advance space politics and policy in the study areas and the continent with various theoretical perspectives with different forms of data and their sources, research methods and techniques. Amidst these, more quantitatively-driven research is required in assessing the contribution of space policy programmes to different but specific development and security goals. Accordingly, specific research efforts need to be devoted to limited indicators of space policy and attendant development and security goals to allow for better and more exhaustive case by case interrogations. Besides, universities in the study areas and the region should commission and sponsor relevant databases as well as subscribe to existing ones to make such data available for academic interrogation and enhance African narratives by Africans in space policy, politics, development and security debates.

Future research is encouraged to devote more attention to the extent that other African countries perceive and reference Nigeria and South Africa as models in space policy and politics in the continent. More attention is also required to raise the prospects and minimise the challenges of space policy in the study areas and beyond. In this case, the important of space popularisation cannot be overestimated in advancing academic and policy interests. There is a need to vote more research attention to bridge the gap between academic community and the population, from where decision-makers with little or no knowledge of the subject matter are drawn. Space S&T should be promoted among other STEM educations, with emphasis on capacity building and quality of education that is available in traditional fields of pure and applied sciences. At the same time, non-traditional fields such as art, humanity, management, and social sciences need to be revised to promote space popularisation and STEM educations. Amidst these, astropolitics, astro-sociology, space policy studies, space commerce and economics, among other fields deserve better attention to building space-fearing societies in the study areas and the region generally. For a sustainable space programme, it is important for a nation to prepare and develop its population to support the policy in question and the commitments to realise its development and security potentials.

At the national level, there is a need to revisit the priority placed on space policy, programmes and institutions in development and security planning. There is a need to maximise utility of available space capabilities to their fullest potentials to advance the development and security aspirations of the peoples and governments of the affected countries. More funding and prudence management of resources are essential in this case. Besides, more policy efforts are required to further integrate the space sector with other sectors of the economy to ensure interconnectivity that can sustain the industry. Commercialisation of some aspects of space programmes are equally essential to boost funding and sustenance of the sector. At the same time, political will, openness, transparency and innovation and general state institutional capacity building are essential to optimise the development and security potentials of space policy, strategy, programmes, institutions, investments, and capabilities in Nigeria and South Africa, and beyond.

Nigeria and South Africa need to take cautious policy measures to strengthen relationship with each other and other African countries. This is important to maintain their mutual confidence and trusts as well as partnership, which are essential to peace, political stability, socioeconomic development, and security in Africa. In this case, space is another frontier for the two countries and others in the region to display their commitments to leadership and mutual respects with the spirits of Pan-Africanism, with emphasis on brotherhood, collaboration, and regionalism. Their space policy and programmes cannot be disconnected from their foreign policies in this case. As a consequence, these countries need to take the lead in supporting other African countries and advocating for international cooperation that will benefits the smaller economies in the region. While this call for some sacrifices on their parts, it does not necessarily need to be flamboyant and wasteful display of space power. Besides, as models for many countries in the region, these countries need to manage their competitive advantages, strengths, and capabilities to minimise its harmful effects and challenges internally and internationally. Finally, these and other countries need to support the narratives of African experiences and perspectives by Africans, including in space politics and policy, to reduce the underrepresentation and marginalisation of the region in the global map of knowledge production in this and related fields of study. In these processes, the international communities need to support these countries among others in the region to ensure globalisation of knowledge as well as shared global prosperity and human security.

References

Abiodun, A.A. 2008. Future trends in the global arena: Challenges and opportunities for Africa. *African Skies* 12: 12–22.

Abiodun, A.A. 2017. *Nigeria's space journey: Understanding its past, reshaping its future*. Abuja: African Space Foundation.

Akinyede, J.O., and R.A. Boroffice. 2013. *Nigeria's quest in space*. Ile-Ife: Obafemi Awolowo University Press.

Froehlich, A. 2019. *Embedding space in African society: The united nations sustainable development goals 2030 supported by space applications*. Cham: Springer.

Froehlich, A., and A. Siebrits. 2019. *Space supporting Africa: A primary needs approach and Africa's emerging space middle powers*. Cham: Springer.

Froehlich, A., N. Ringas, and J. Wilson. 2020. *Space supporting Africa: Security, peace, and development through efficient governance supported by space applications*. Cham: Springer.

Gottschalk, K. 2010. South Africa's space program. *Astropolitics* 8 (1): 35–48.

Harding, R.C. 2013. *Space policy in developing countries: The search for security and development on the final frontier*. New York: Routledge.

Helfrich, K. 2014. Defence minister enters the spy satellite fray. *DefenceWeb*, January 23.

Joubert, J. 2014. Scandal of R1.2 billion missing SA satellite. *Time*, January 19.

Martinez, P. 2016. The development and initial implementation of South Africa's national space policy. *Space Policy* 37: 30–38.

Munsami, V. 2014. South Africa's national space policy: The dawn of a new space era. *Space Policy* 30: 1–6.

Ogunnubi, O., and S. Oyewole. 2020. Africanism, power politics and international relations in Africa. In *Power politics in Africa: Nigeria and South Africa in comparative perspective*, ed. O. Ogunnubi and S. Oyewole, S. Newcastle: Cambridge Scholar Publishing.

Oyewole, S. 2017. Space research and development in Africa. *Astropolitics* 15 (2): 185–208.

Oyewole, S. 2020a. The quest for space capabilities and military security in Africa. *South African Journal of International Affairs* 27 (2): 147–172.

Oyewole, S. 2020c. Civil-military relations: Conflict and cooperation between military bases and host communities in Nigeria. *African Security* 13 (4): 353–379.

Oyewole, S. 2023b. The African orbital journey in space age: From spectator to participant. *Vanguardia Dossier* 88 (15): 72–75.

Oyewole, S. 2020b. Military capabilities of regional powers in Africa: Nigeria and South Africa in comparative perspective. In *Power politics in Africa: Nigeria and South Africa in comparative perspective.*, ed. Olusola Ogunnubi and Samuel Oyewole. Newcastle: Cambridge Scholar Publishing Ltd.

Oyewole, S. 2021. Air power capabilities and applications of African regional powers. Presentation at the (British) Royal Air Force Museum Conference on "New Thinking in Air Power", September 16.

Oyewole, S. 2022a. How space research can help Africa meet the SDGs. Presentation at a webinar of the University of Johannesburg's Institute for Pan-African Thought and Conversation (IPATC) on "Reaching for the final frontier: The African Space Sector", September 29.

Oyewole, S. 2022b. Air and space strategies in African disaster management. Virtual Lecture of the (British) Royal Air Force (RAF) Museum on August 11.

Oyewole, S. 2023a. Military space strategies and African realism: Egypt, South Africa and Nigeria. In *Oxford handbook of space security*, ed. S. M. Pekkanen and P. J. Blount. Oxford: Oxford University Press.

Porta, D. 2008. Comparative analysis: Case-oriented versus variable-oriented research. In *Approaches and methodologies in the social sciences: A pluralist perspective*, ed. Donatella della Porta and Michael Keating. Cambridge: Cambridge University Press.

Smith, K. 2015. Africa as an Agent of International Relations Knowledge," Cornelissen, S., F. Cheru and TM. Shaw (eds.)., *Africa and International Relations in the 21st Century*. New York: Palgrave Macmillan.

van Wyk, J.A. 2009. South Africa's space policy and interests: A New dawn or a black hole? *Strategic Review for Southern Africa* 31 (2): 46–73.